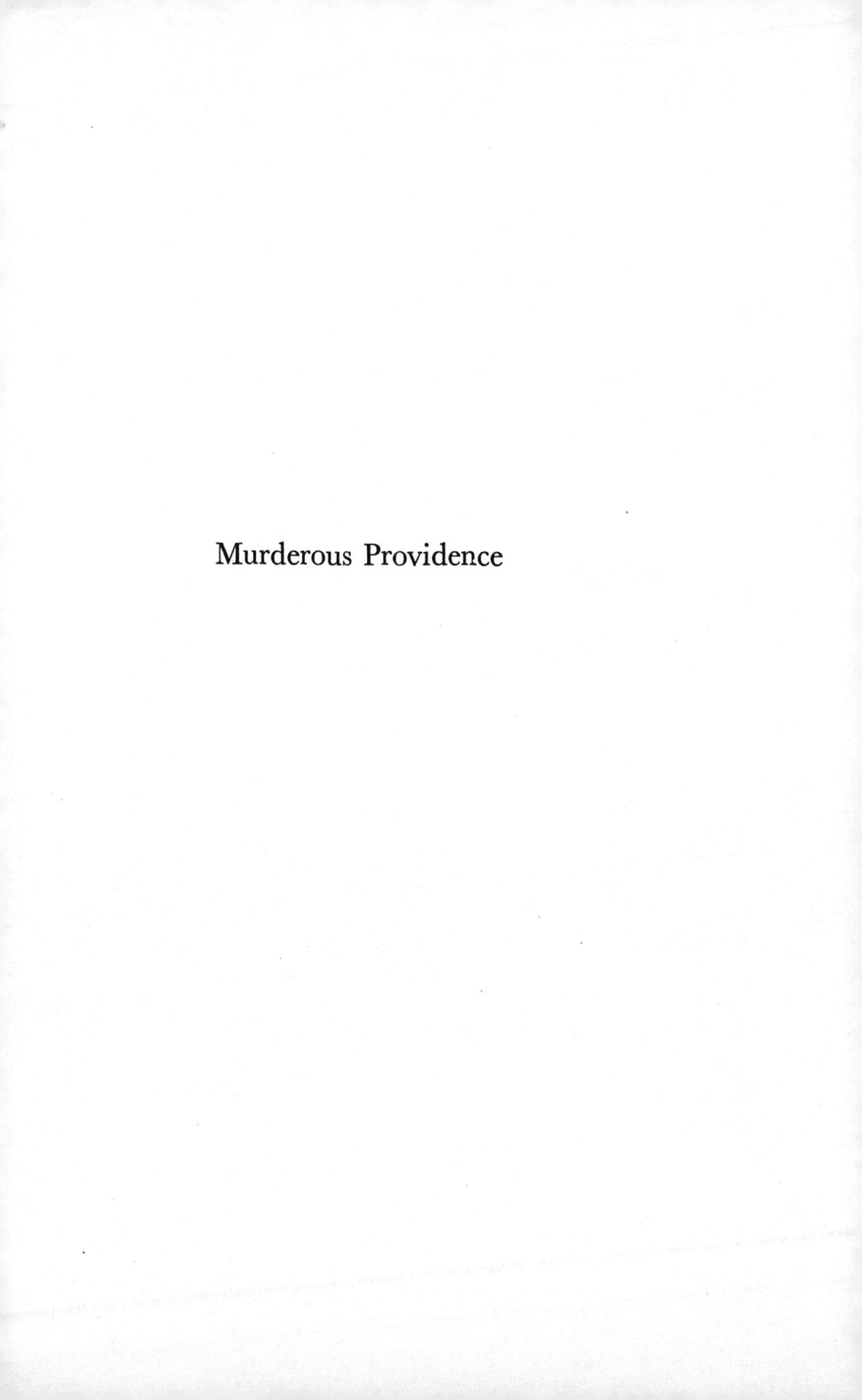

Murderous Providence

Murderous Providence

A Study of Pollution in Industrial Societies

Harry Rothman

The Bobbs-Merrill Company, Inc.
Indianapolis · New York

TD
174
.R72
1972

The Bobbs-Merrill Company, Inc.
Publishers/Indianapolis/New York

Copyright © 1972 by Harry Rothman
First published 1972 by Rupert Hart-Davis Ltd., London

All rights reserved. No part of this publication
may be reproduced, stored in a retrieval system,
or transmitted, in any form or by any means,
electronic, mechanical, photocopying, recording,
or otherwise without the prior permission of the
publishers.

Library of Congress catalogue card number: 77-173222
Printed in the United States of America

Contents

Introduction ix

Part 1
The Genesis of Pollution Society

1 The Ecosystem 3
2 Population Dynamics and the Balance of Nature 14
3 Man, the Unique Life-Form 20
4 From Hunting and Gathering to Industrial Production 32

Part 2
The Current Situation

5 The Pollution Crisis in the Cities 45
6 The Factory and Place of Work 71
7 The Rural Environment 90
8 Fresh Water Pollution 113
9 The Minamata Disease and Methyl Mercury Poisoning 135

10	Industry and Pollution—	
	The Automobile Industry	148
11	The Detergents Industry	159
12	The Tourist Industry	174
13	Military Technique and Pollution	182
14	Global Pollution	204
15	Pollution of the Marine Environment	218
16	Oil Pollution and the Future Industrialisation of the Sea	237
17	Pollution in the Soviet Union	248

Part 3
The Techniques of Pollution Control

18	Pollution Control Technology	259
19	Legislative Control of Pollution	274

Part 4
The Ultimate Solution—Social Reform or Revolution

20	Pollution and Political Priorities	293
21	Economic Rationality	309
22	Pollution, Romanticism and Ideology	323
23	Hope for the Future	334
	Notes	339
	Bibliography	355
	Index	367

Acknowledgements

I have attempted to synthesise information from many disciplines and I am only too aware of the enormous debt that I owe to the original researches and ideas of other people. I therefore thank all those individuals and organisations, far too numerous to list separately, who gave me articles, reports and other information. I only trust that in my attempt to relate their material to a wider conceptual scheme I have done them justice.

I received a great deal of help and encouragement from colleagues at Manchester University. In particular I thank Dr Norman Lee and his colleagues at the Manchester University Pollution Research Unit for allowing me to attend an invaluable series of seminars during the autumn and winter 1970-1971. Professor F. R. Jevons and members of my department showed understanding and forbearance when the writing pressure was greatest.

I should like to record my gratitude to Joe Gilgan for his indefatigable efforts on my behalf. For help in research, typing and various other forms of personal assistance I thank Anne Berg, Neal Smith, Kenneth Green, Lynn Walsh, Dave Aron, Catherine Quinn, Anita Ray, Margaret Bruce, Edna Foster, Donald Cooper, Rodney Campbell, Ian Gough, Paul Arnold, Walter Ryder, Göran Löfroth, Stanley Metcalfe and Ivan Sorah.

Introduction

*A murderous Providence! A Creation that
groans, living on Death
Where Fish & Bird & Beast & Man & Tree & Metal & Stone
Live by Devouring, going into Eternal Death
continually!*
                                                William Blake.

Mankind is confronted by a multi-faceted crisis of society of unprecedented proportions. The social crises which man has continually faced throughout history have taken a qualitative turn for the worse; whereas formerly such crises were localised, they now involve the whole planet. Environmental pollution, which carries mankind's problems into the very structure of Nature itself, is a major part of a general social crisis which must be overcome if civilisation is not to relapse into barbarism.

It is ironical that the development of industry should create an environmental crisis, for technology gave mankind sufficient control over his surroundings to enable him to raise himself above the level of the beasts. Yet by our apparently increasing control over the environment, we may have sown the seeds of a complete loss of control.

Why should this have occurred? Is it something inherent in technology itself? Or is the source of our crisis to be found in the nature of the social and economic organisation of the society which utilises technology? Is there any hope for civilisation? These are the questions which I set out to answer.

My major thesis is that pollution is a function of our society's mode of production, the historically conditioned fashion in which society obtains the necessities of life. Pollution is a social and economic problem which takes on the appearance of a technical problem. Thus all attempts to find a 'technical' solution to the pollution problem are doomed to failure because they merely treat effects rather than causes. The causes result from social contradictions within the very basis of society.

Capitalist society made more rapid progress than previous social forms because it was the first society able to break with traditionalism, the first society to base its activities on the maxim that nothing is worth doing unless the economic benefits outweigh the costs of doing it. Such a view is an enormous advance on the traditionalist view that if a thing has been done in the past carry on with it and if it has not been done before don't try it. The capitalist form of society encouraged mankind to develop economic rationality and this made for a terrific rate of progress. Unfortunately this rationality took on a distorted form because capitalist society is organised as though the rational pursuit of private ends by separate individuals and enterprises will result in what is best for the whole society. This is manifestly not the case and we find ourselves continually doing things which are good for ourselves but bad for other people. Pollution is a particularly nasty result of such social irrationality.

I argue that the *private* economic rationality of the profit-seeking business enterprise is a *murderous providence* because it cannot guarantee the optimum use of resources from the point of view of society as a whole. It cannot avoid continually creating situations which cause the pollution of our environment. If pollution is to be cured and mankind survive in a healthy world the individual enterprises of society have to be co-ordinated by a common goal directing the activity of the whole society. Only by creating such a society, which governs its actions by a social instead of private economic rationality, can mankind live in an affluent but non-polluting culture.

PART ONE

The Genesis of Pollution Society

I
The Ecosystem

The Sun and Life

The heart of life is a hydrogen bomb. Without the sun there would be no life. Throughout history men have recognised this link and worshipped the sun. Modern man's fetish for sun-bathing is a symbolic charging up of vital batteries depleted by the stress of urban life, a search for the optimum route for vital force.

The temperature and composition of the sun are such that hydrogen is transmuted to helium releasing enormous amounts of energy in the form of electromagnetic radiation. This radiation ranges from the short-waved X-rays and gamma rays, to the long-waved radio waves. About half of the energy is in the visible spectrum, i.e. light.

Life feeds on the energy contained in light. Many of the basic problems of life can be resolved into the problem of energy availability, the problem of taking solar energy into living organisms and the passage of energy from one form of life to another. The evolutionary process can be conceived of as a self-optimising system of energy flows in response to changing environmental conditions. Human progress is the emerging awareness of the possibility of the conscious control of this energy flow to mankind.

The continuous nuclear fusion of the sun pours energy into space. But most of it is dissipated; only a fifty-millionth of it ever reaches the earth.[1] This continuous output is termed the 'solar flux', and after other losses, only $\frac{1}{10}$ of a fifty-millionth part of it ever reaches the life of this planet. And yet by human standards, this amount is enormous, about two calories per square centimetre per minute. Put in terms of an electricity bill, if we paid the very low price of one cent per kilowatt-hour for solar energy, the bill would be 478 million dollars per second.

The main reason for the high rate of solar energy loss is the reflecting properties of the earth. Dr Rudolf Geiger calculated that 42% of the incoming radiation is reflected back into space. Clouds and dust are chiefly responsible. Another 10% of the solar flux is absorbed by ozone, oxygen, water-vapour and carbon dioxide. As we shall see, some forms of pollution may already be interfering with these processes.

The only life-form capable of absorbing its energy directly from the sun's rays is green-leafed plants. All other life-forms must absorb it indirectly in the form of food based ultimately on the nutrient properties of plants. The process by which energy is thus transferred from one life-form to another has been built up slowly over millions of years. Let us just examine how long it has taken for this complex chain of events to be built up.

The earth is 5,000 million years old, and life about 2,500-3,000 million years old, so the current energy balance of the earth has taken thousands of millions of years to evolve. Man, in 200 years of ever-increasing industrialisation, is threatening the ancient régime. Let us try to visualise the sharp and sudden impact of industrial society on the living world.

Imagine that the world is seven days old, the solar system and the earth formed on Sunday morning. The first life would then appear some time on Wednesday. For most of the remaining week we would see very little that is familiar. Then early on Saturday morning we would see the first vertebrates and fish. By mid-day we would see the first creatures leaving the sea and conquering the land. By six o'clock we would hear the first faint hum of insects, and by seven, the first flowering plants would begin to bloom. Also by seven o'clock, the heavy feet of dinosaurs would be thrashing through the undergrowth, but not till ten o'clock would we come across our first familiar mammals. The whole of man's existence would be the last five minutes of Saturday. The first city would rise in the last half-second, and the whole of recorded history would happen in the last quarter-second. If you were so unfortunate as to blink in the last dying moment of Saturday, you would miss the rise of industrial society. And yet this instant of time is so sudden, so ruthless, so abrupt that it may be shaking the very foundations of life on this planet.

The envelope of life that films our planet is termed the biosphere. This term was coined by the Austrian geologist Eduard Suess in 1875. He conceived of the biosphere as being one of several planetary envelopes — the atmosphere, the hydrosphere and the earth's rocky crust or lithosphere. The biosphere concept was popularised among biologists by the Russian Vladimir Vernadsky in the nineteen-twenties.[2]

The exact delineation of the biosphere is difficult. There is more of it in some places than others, more of it in a tropical rain forest for instance than in a desert. It is the living portion of the earth, a thin outer layer, the product of a continual interaction between living and non-living substances. It can extend some way into the atmosphere, into ocean depths and even underground into caves. Basing itself on the photosynthetic activity of green plants, the biosphere is able to collect, store and distribute a part of the energy poured on to the earth by the sun.

As the biosphere evolved and developed, its activities radically altered the earth's structure. It laid the basis for sedimentary rocks such as limestone, and it created the oxygen-laden atmosphere so necessary for modern life-forms. We should, says Professor V. Kovda, consider the biosphere to be 'an ancient, extremely complex, multiple, all-planetary, thermo-dynamically open, self-controlling system of living matter and dead substances, which accumulates and redistributes immense resources of energy and determines the composition and dynamics of the earth's crust, atmosphere and hydrosphere'.[3]

The biosphere maintains a stability and dynamism over time because living organisms evolve. The evolutionary process of natural selection enables living organisms to adapt to the changes which they bring about by their ceaseless inter-action with each other and the physical environment. Our concern in this book is to discuss how man can best adapt his behaviour so as to avert the potentially damaging effects brought about by his industrial civilisation.

Without the rational planning of society based on a scientific understanding of the laws of nature, humanity will not be able to make the necessary adaptations. Science has a vital role to play in this task. Unfortunately, scientists cannot always offer a unified view of reality to correspond with the unity of nature. Their thinking is fragmented into different scientific disciplines. This specialisation is a necessary division of labour, but it poses great problems when considering the impact of human activities on the biosphere.

Unity and Specialisation in Science: Ecology

Scientific disciplines tend to correspond to levels of organisation of matter, e.g. the physical sciences correspond to the inorganic level, the biological sciences to the organic level, the behavioural sciences to the psychic level, and the social and political sciences to the socio-cultural level. Within each of these broad divisions, and even between them, there are sub-levels, e.g. in the inorganic level physics operates on the atomic and energy levels whereas chemistry concentrates on the molecular level. (Within physics there are further subdivisions e.g., the physics of subatomic particles.) There are of course growing breeds of scientists who work in the transitional areas between levels, such as the molecular biologists, who are examining how the complexities of large molecules relate to the vital property of the organic level.

The complexity of nature is such, that no scientist could hope to traverse all of it. The scientist has had to specialise. For instance in the study of living matter alone, there is the level of the biologically active macromolecule, such as DNA or the viruses; there is the second level of the cells; cells are organised into tissues, and the tissues form organisms; organisms are organised into populations which are integrated with one another into ecosystems. The highest organisation of life is the biosphere.

Each level here requires detailed and specialised study, but the danger is that scientists tend to think and predict within their own particular level. The cytologist sticks to the cell, the histologist to tissue-structure and so on. Each speciality evolves its own 'paradigm', that is a set of theories, methods and problems which aid its research but also tend to emphasise its separateness. This leads to the possibility of major errors of judgment when it comes to making decisions that cut across these levels. One such area is that of environmental pollution. Who is the expert? For example, in cases of harmful effects of pesticides, the medical man is often brought in as the expert. Thus, on the basis of their discipline, medical men, finding no direct effect on human health, have pronounced as safe compounds which produce ecological havoc.

However there is within science also a corrective tendency towards unity. Thomas Kuhn, in his *Structure of Scientific Revolutions*,[4] shows how 'normal science' is thrown into a crisis from time to time by the mass of new data that cannot be handled within the accepted paradigm. Apparent anomalies and contradictions may force separate specialities to be thrown together, and a new paradigm is developed, more appropriate to a heightened awareness of nature.

There is need for such a revolution at the present time. The enormous potential for change in modern technology makes it imperative that a scientific discipline competent at the biosphere level be developed. Gross effects on the biosphere, e.g. the 'greenhouse' effect, are the subject of speculation rather than science. Ecology is the scientific discipline most germane to the development of a science of the biosphere.

Ecology was first spoken of in the latter half of the nineteenth century. In 1870 the German biologist Ernst Haeckel wrote: 'By ecology we mean the body of knowledge concerning *the economy of nature* ... the study of all the complex *interrelations* referred to by Darwin as the condition for the struggle for existence.' Charles Elton, who pioneered modern ecological studies, called ecology 'scientific natural history ... (the) sociology and economics of animals'. Edward J. Kormondy says 'the substance of ecology is found in the myriad of abiotic and biotic mechanisms and interrelations involved in moving energy and nutrients, regulating populations and community structure and dynamics'. Although ecology lays stress on interrelation and economy it is, however, a mistake to see this as a scientific validation of the belief in the *'natural balance'*. Rather, ecology will lay the basis for making more rational our attempts to change the existing interrelations of life-forms so as to optimise the energy and materials flow to our own species.

Ecosystems

The structural and functional units which ecology studies are ecosystems.[5] An ecosystem has been defined as 'the invisible complex in which no animal or plant exists in total isolation, and no factor, physical or biotic, operates in complete independence'.[6] This is an all-embracing definition but the ecosystem is hard to pin down. The chemist can isolate his molecule prior to analysing it, the cytologist can fairly easily obtain a cell, pull it to bits ... and now even put it together in an endeavour to understand how it works. Unfortunately for the ecologist, ecosystems are less definite and before starting his research the ecologist has to be clear which level of ecosystem he intends to study.

Some ecologists have spent many years researching 'microhabitats', the ecosystem of, say, a cow-pat, others have done equally sound work in 'macrohabitats', such as the ecosystem of a large forest or prairie. We have therefore to conceive of a whole interacting hierarchy of ecosystems from microecosystems through to macroecosystems. In fact in the broadest sense the whole biosphere can be said to constitute an ecosystem, and we could consider the whole solar system as an ecosystem since the biosphere would be very sensitive to any changes in the present relation between the earth and the sun. But, generally speaking, a more limited definition has been used by the research ecologist in which the dimension of an ecosystem may range between a few feet and many miles in length and from inches to many feet in height.

The Origin of the Ecosystem Concept

In spite of such apparent looseness the ecosystem has proved to be a most valuable operational concept for the ecologist. All ecosystems have certain general structural and functional characteristics.

The concept of the ecosystem enabled the ecologists to begin to escape from the strait-jacket imposed by the historical division of biology between botanists who studied plants, and zoologists who studied animals. Much of the early work done by plant ecologists classified the living world in terms of 'plant communities', and the animal ecologists in turn classified the living world of 'animal communities'. Much valuable work was done by both groups, but by the 1930s it was obvious that the division was leading to a great deal of confusion, for 'plant communities' and 'animal communities' were tending to be seen as separate entities with some 'links' between them. Futhermore, they both underrated the non-living components of the environment. The ecologists had turned the real world into a refracted image of their own academic world which was divided into botanical and zoological communities.

The necessity of a break with the traditional approach was argued by the Oxford ecologist A. G. Tansley, in an article called 'The use and abuse of vegetational concepts and terms'.[7] His case for the ecosystem concept is worth quoting at length.

'What we have to deal with is a *system*, of which piants and animals are components, though not the only components.' Further, '... The more fundamental conception is, as it seems to me, the whole *system* (in the sense of physics), including not only the organism-complex, but also the whole complex of physical factors forming what we call the environment of the biome (a living community of plants or animals in a particular area or zone)—the habitat factors in the widest sense. Though the organisms may claim our primary interest, when we are trying to think fundamentally we cannot separate them from their special environment, with which they form one physical system ...

'... the inorganic "factors" are also parts—there could be no systems without them, and there is constant interchange of the most various kinds within each system, not only between the organisms but between the organic and the inorganic. These *ecosystems*, as we may call them, are of the most various kinds and sizes. They form one category of the multitudinous physical systems of the universe, which range from the universe as a whole down to the atom.'

Tansley also recognised that the size and extent of ecosystems would be determined by the purposes and methods of particular studies. This is important in Lindeman's definition of the ecosystem as 'the system composed of physical-chemical-biological processes active within a space-time unit of any magnitude'.[8] Lindeman's researches, packed into a tragically short life, exerted a powerful influence on modern ecology. He was one of the pioneers of a research approach which conceived of ecosystems as systems through which energy flows and material cycles. This approach has dominated modern ecology and it is particularly relevant to any rational approach to the pollution problem. However, bearing in mind our previous discussion of the operational nature of ecosystems, a word of warning is required about the tendency to a simplistic application of the ecosystem concept to problems facing society. The ideal of a 'steady-state' society with complete recycling and no waste is a long way off and requires, besides massive advances in ecological science, a total restructuring of society. The view which defines ecosystems as characteristically closed with complete cycling of nutrients such as nitrogen, phosphorus, etc., is a convenient *abstraction*. Unfortunately, reality itself is more complex, 'cycles are always incomplete and ecosystems are never quite closed systems'.[9] Each ecosystem receives and supplies energy and matter from other systems. How all these ecosystems within ecosystems function within the biosphere is still, despite a great deal of effort, only vaguely understood. For this reason it is essential to guard against the temptation to see ecology as the miracle science which will with a wave of its scientific wand clean up the mess created by industrial society. For some time to come ecology will tend to pose the problems to the 'planners' rather than provide the formulae for easy solutions, e.g. if society continues to burn fossil fuels on the present scale there *may be* disastrous effects on world climate.

The Flow of Energy and the Cycle of Matter

Energy from the sun is trapped by 'producers', the green plants which by means of a complex biochemical process called photosynthesis are able to synthesise biological materials from the energy of the sun and chemicals from the air and soil. The plants are eaten by certain kinds of animals, 'herbivores', which in turn are eaten by 'carnivores'. The dead bodies of these three types of organisms form food for a myriad of organisms called 'decomposers'. The role of the 'decomposers' is a vital one, for they release most of the basic elements bound up in dead bodies making them available for re-use by the producers. A 'fertile' soil depends on the availability of a healthy community of such decomposers. Thus whereas the energy flows only one way through the system and is lost at the end, the nutrients are rotated and cycled. The general principles of water, carbon, nitrogen cycles are described in elementary biology text books; there are also cycles of phosphorus, sulphur and various metals, the full implications of which are still a subject for scientific speculation.[10]

Like everything else ecosystems conform to the basic laws of thermodynamics which tell us that firstly, energy cannot be created or destroyed but only transformed from one form to another, e.g. chemical energy to heat energy. Secondly, systems tend towards a state of entropy or disorganisation. This means therefore that ecosystems must have a continual source of energy to survive and maintain their complex organisation, and also as energy flows through the ecosystems more and more of it is transformed into a non-usable form (heat) and is dissipated.

The 'thermal efficiency' of ecosystems is very low. That means that they use only part of the energy available to them. Firstly, as we have noted, plants only trap about 1% of the sun's energy that reaches the ground. Of course, photosynthesis can only use part of the light spectrum. However, even if one measures their efficiency as a percentage of the *usable* light it still only rises to 2–6%. Secondly, at each link in the food chain or web there is another energy loss. This has led the ecologists to the concept of 'trophic levels'. These levels correspond to the links in the chain: producer–herbivore–carnivore. Roughly speaking only about 10% of the energy held in a lower trophic level is transferable to the next. We can imagine a 'food pyramid' in which a given weight of soil supports a smaller amount of grass which in turn will carry a smaller weight of cows which in turn supports an even smaller weight of carnivores. At each level energy has been lost as heat or used up by growth, respiration and other self-maintenance processes necessary to keep alive the organisms composing that level.

From this simple model of energy and material flows in ecosystems we can see several general principles of importance to a more rational planning of society. Since men cannot photosynthesise, man belongs in the higher trophic levels, as a herbivore *and* carnivore. An optimum production of food demands that we understand how to improve the efficiency and output of 'producers', i.e. crops; that we seek to improve the efficiency of energy flow through man-oriented ecosystems. For example, more food is available to mankind for less expenditure of energy if we keep to a vegetarian diet. If the world wants the maximum possible number of people then they will have to eat plants directly instead of meat. Finally, the introduction of any toxic substances into the environment has to be avoided for they can easily enter our food chains as a result of the various natural cycling processes in ecosystems. Later chapters will show at length the havoc such pollution is playing in the human ecosystem.

Our simple model ecosystem ignores a special type of ecosystem which is not *directly* dependent on solar radiation. These are 'detritus-based' ecosystems which depend, in the main, on releasing energy locked in organic material produced in other ecosystems. The energy relations of a small spring, Root Spring at Concord, Massachusetts, were studied in great detail. It was found that three-quarters of the energy of this ecosystem was obtained from dead organic material or detritus which flowed into it from other systems. Another classical form of detritus-based ecosystem is forest litter, the accumulation of dead leaves, etc., found at the foot of forest trees.

The detritus-based ecosystem is not simply a matter of academic interest. Mankind can be said to have based the industrial revolution on a sort of detritus-based ecosystem. Fossil fuels such as coal and oil are the detritus of ecosystems which died millions of years ago in the Carboniferous era. They have provided the primitive accumulation of energy capital necessary to raise the productive forces of mankind to a new level. However, it is clear that our survival as members of a detritus-based ecosystem can only be a temporary one. Today this energy capital is being ruthlessly squandered and, at the same time, the resulting pollution is reducing the efficiency of the ecosystem components upon which depend the health of both man and the biosphere itself.

2

Population Dynamics and the Balance of Nature

Animals rarely increase in numbers to a level at which further increase is prevented by starvation. One of the unsolved questions of ecology is what limits the number of individuals of any given species? The branch of ecology which seeks to uncover the laws of population control is called population dynamics. The number of individuals of a given species in a particular place changes over time. In theory their numbers fluctuate between nearly zero and the ultimate capacity of the habitat. But what is 'ultimate capacity'? One can *imagine* a population rising to a level beyond which any further increase in numbers would lead to its extinction because the habitat's resources become exhausted or destroyed—however, in reality this never happens. Species generally become extinct because their fecundity has been too low or they are struck by a catastrophe.

There are two types of environmental factor which may alter

population sizes. Those which act on the population no matter what its density, are called *density-independent* factors. The most important of these is weather, e.g. a hard frost may hit a low population as hard as it hits a high population. The second type are called *density-dependent* factors. Their effect on a population changes according to the density of population. As the population density begins to increase so their influence increases. Such control factors include other species competing for the same resources (interspecific competition), predators, parasites, disease pathogens, and competition between members of the same species (intraspecific competition). For example, the higher a population density the greater the competition will be for the available food supply, nesting sites or any other resources required by competing individuals or species. It is generally agreed that the control of numbers of any species is due to the combined action of density-independent and density-dependent environmental factors. However their relative importance is a matter of great controversy. For example, some workers believe that only density-independent factors prevent a population dying out and that whether or not a species' population increases depends on the combined action of density-independent and density-dependent factors. Intraspecific competition is thought by some workers to be a relatively rare form of density-dependent control which only occurs when a population approaches the environmental capacity of its habitat. It is obviously an extremely complicated process and, given the present state of research, it is premature to expect any consensus about a general theory of population control.

Some of the most interesting research results concern the effect of changes in population density on behaviour. A pioneering piece of research of this nature was done over 40 years ago on the desert locust. There are two forms of desert locust, a greenish solitary form which looks like a grasshopper and a brownish gregarious form. For a long time entomologists thought that they were two completely different species until one experimenter[1] found that by crowding individuals of the 'solitary' form together he could create the 'migratory' form. In nature the solitary grasshopper-like locust stays in that form until it reaches a certain population density, from which point on more and more of the succeeding generation develop into the larger winged migratory form. These then eventually fly off in great swarms and so prevent the original breeding grounds from being destroyed.

The suicidal migrations of the lemming, a small rodent of Northern Europe, are well known. These migrations seem to occur in three or four yearly cycles. When populations rise beyond a certain point most members develop an urge to travel downhill; as a result great hordes of them eventually reach the sea and plunging into it are mostly drowned. This apparently purposeless behaviour is said to serve the function of preventing their habitat from being destroyed by excessive population.[2]

In other species one finds complex and apparently pointless territorial behaviour and this is thought to be a device to keep down population density. Laboratory studies on rats and mice have shown that overcrowding can lead to stresses which reduce fertility and lessen the survival ability of individuals.[3]

It seems likely that behavioural mechanisms will be found to control population density more often than is usually thought, but we should not jump to the conclusion that this work will tell us how to control human populations. Even in population dynamics the proper study of mankind is *man*. Certainly a growing world population is putting great strains on existing resources, but before accepting the glib talk about 'people pollution' we must fight to remove those social and class privileges which everywhere stand in the way of an equitable distribution of resources.

The idea that there is a 'balance of nature' holds considerable sway. This idea is that the numbers of individuals and species within an ecosystem remains relatively constant, they are in a state of 'harmonious adjustment', 'dynamic equilibrium', or in more crude terms of 'balance'. Any drastic changes are a result of 'upsetting the natural balance'.

A more sophisticated view of 'balance' has been put forward by Professor A. J. Milne.[4] He believes that the balance is not between the population and the environment but in the environment itself. He points out the continued existence of a species depends on favourable conditions, those allowing a population increase being 'balanced' in their incidence and intensity by unfavourable conditions, those causing a population reduction. This balance is achieved by a two-fold mechanism. Firstly there is generally a natural limit to the severity of unfavourable conditions which prevents them reducing a population to extinction. Climatic conditions generally vary between certain seasonal limits; we are, for example, unlikely to suddenly get a prolonged frost in a region of tropical rain forests. Secondly, favourable conditions are turned into their opposite by the population itself before its numbers surpass the environmental capacity.

However, equilibrium and steady states in nature are only relative; most of the species that have ever existed are extinct today. A species will eventually meet a set of environmental conditions severe enough to reduce its numbers to zero. The 'balance of nature' is both dynamic and relative and not static. Any particular ecosystem is but a temporary feature; it grows, matures, enters a period of senescence and eventually dies. Sometimes ecosystems may rise again, sometimes they are replaced for ever like the ecosystems of the Carboniferous Era. The chief difference in the life-cycle of an ecosystem and that of our own is the time scale. A man's life-cycle takes less than a century, whereas the life-cycle of an ecosystem may be hundreds or even thousands of years.

The interaction between the living and non-living components of an ecosystem provides an inner dynamic. An ecosystem can change itself, the very activity of a particular species ultimately changes the physical conditions which initially favoured it. We must remember that the activities of living organisms have over millions of years completely reshaped the surface of the earth. Life has released oxygen to the atmosphere, and has concentrated and shifted minerals around the planet e.g. limestone, oil and coal. Of course, these processes took millions of years but they were a result of the smaller-scale, more rapid processes occurring within ecosystems. As time progresses, later generations of the original species of a particular ecosystem find that its changing conditions are no longer suitable and they either migrate or die out. At the same time a new community of plant and animal species is developing within the original one. A new set of relationships is established which in turn alters the area yet again, setting the scene for a further 'communal succession'. Eventually a more stable situation or climax community may develop. Climax communities are relatively stable, the tempo of community change having slowed down. Particular climatic regions have characteristic climax communities, e.g. in the South-Eastern United States the climax is the oak-hickory wood, in the area from the Mississippi Valley westward to the Rockies it is a prairie, in the far north of North America and Asia it is the tundra. The climax could be considered the naturally optimum system for a particular set of climatic conditions. But even here drastic climatic and geological changes and nowadays man-made changes will eventually push the system out of 'balance'.[5]

A gradual process of silting up changes the type of community in the region originally covered by a lake. The climate determines the eventual climax; in a dry region marshes develop into prairie and in wetter areas into a forest.

Man increasingly finds himself interfering in the natural process of succession and struggling to maintain a 'balance' of his own choice; this is what agriculture is all about.

Human agricultural activities have so far aimed to produce simplified ecosystems containing as few species as possible. Any species which interferes with the flow of energy from corn to man is regarded as a pest and has to be eliminated or reduced in numbers. Man seeks the highest possible share of the energy trapped by the producers. It has been argued[6] that the main danger with such simplified ecosystems is that they are liable to be easily disturbed with the result that sudden population fluctuations occur, in other words the crop fails. Populations are said to be less liable to fluctuations in complex climax communities such as a tropical rain forest. There the great range of different species gives a complexity of relationship which seems to buffer the community against change. However, precisely because of their simplicity the agricultural ecosystems are extremely vulnerable to attack by pests and disease. In order to control pest populations, increasing amounts of chemical pesticides, which can become extremely dangerous pollutants, have to be used. We can see therefore that the concept of the optimum energy flow in the human ecosystem may not necessarily be that currently practised by agriculture. Present biological and economic analysis seem to contradict one another. Ecologists may argue for diversity but agricultural economics argues for simplicity. In Britain the diversity afforded to agriculture by hedgerows is destroyed in the interests of current economic rationality.

The apparent contradiction between ecology and economic rationality is something we shall return to later.

Can we ever hope to control the environment using laws of ecology? The 'technological optimists' will argue that with enough knowledge and 'scientific planning', 'the biosphere can be manipulated as a man-controlled system, which will provide the most favourable conditions for the welfare of mankind. Any manipulative measures must of course take account of the limits of tolerance and plasticity of the biosphere.'[7] I refer to this as 'technological optimism' because it ignores a vital component of the equation—political economy. Scientific knowledge or even the will to plan are not enough if the human environment, which is determined by its political and economic structure, is not favourable. It is not difficult to demonstrate, firstly, that industrial capitalism has no functional conception of the welfare of *mankind* as a whole; secondly, there is no exact scientific understanding yet of the 'limits of tolerance and plasticity of the biosphere'.

3
Man, the Unique Life-Form

An animal which pollutes its own environment must be a rather unusual kind of animal. In the last chapter we saw how each ecosystem, indeed the whole biosphere, has within it a series of complex pathways along which energy flows and matter is constantly recycled. All life-forms adjust to these patterns and to each other. This is a dialectical process in which the living species perpetually struggle to survive. This struggle produces a constant but unconscious development of new life-forms, new relationships and new ecosystems. It was only a matter of time, however, before this process of natural selection evolved the ultimate survival mechanism—consciousness. With the evolution of man, 'nature attains consciousness of itself'.[1]

But with the evolution of this conscious creature, Pandora's box had been opened, releasing forces beyond the control of existing natural patterns. The continued existence of life on this planet now depends upon the conscious actions of man. Having destroyed the old unconscious processes, he must become the new controller; but this means that human consciousness must acquire the dialectical sensitivity of nature itself.

Human consciousness did not emerge in a fully developed form. It is a continually developing product of man's struggle to survive. This struggle is not undertaken by isolated individuals but takes place within the structure of complex social relations. Man's conception of his relationship with his environment is determined by the type of society in which he lives. To understand why mankind pollutes the environment, it is necessary to expose the economic and social relationships that produce a false consciousness blinding men to the consequences of their actions to the environment. The theoretical basis for such an analysis was pioneered by Frederick Engels and Karl Marx.

Frederick Engels' Dialectical View of Man and Nature

Is man meant to serve nature, or nature man? This question poses false alternatives. Today, we see its two aspects personified in the technocrat, priding himself on ruthless machine-like efficiency, laying waste great tracts of land; and, on the other hand, the mystic hippy, who adopts a passive role in society, submitting to cosmic influences. The one stresses man's separateness from nature, and the other entirely relinquishes his active role in events.

Frederick Engels was an early propagator of the dialectical view of nature, which resolves the contradictions between man and nature, and exposes the false alternatives posed above. By a dialectical view of nature, Engels meant one which recognises that '... in nature nothing takes place in isolation. Everything affects every other thing and vice versa'. He also believed that it was not only ignorant laymen who failed to see this but 'because this all-sided motion and interaction is forgotten... our natural scientists are prevented from clearly seeing the simplest things'.

Engels discussed the undialectical attitude to nature of industrialists, whose actions often had unforeseen consequences, such as soil erosion and pollution, adding that nature always takes its revenge on us if we ignore its laws. However, Engels certainly did not think we should subordinate ourselves to nature, though we should recognise the fact that we do not 'rule over nature like a conquerer over a foreign people, like someone standing outside nature—but that we, with flesh, blood and brain, belong to nature, and exist in its midst, and that all our mastery of it consists in the fact that we have the advantage over all other creatures of being able to know and correctly apply its laws.' '... We are learning to understand these laws more correctly, and getting to know both the more immediate and the more remote consequences of our interference with the traditional course of nature.' '... We are more and more placed in a position where we can get to know, and hence to control, even the more remote natural consequences at least of our most ordinary productive activities.'

Engels emphasised that mankind had been slow to understand the remote consequences of his actions on nature, and that the understanding of the remote social consequences of changes in production had been even slower. He cited the fact that the inventors of the steam engine had no inkling of the social revolution that it would create. Equally, the creators of the 'Green Revolution' today in South-East Asia misjudged the consequences of their actions. The 'Green Revolution' was the term given to the introduction of hybrid rice, giving far greater yields than traditional varieties. It was hoped that famine and social unrest could be avoided in this way. In fact social revolution is probably being brought even closer by this technical development. The new agricultural techniques involved in harvesting the hybrid rice, more fertilisers, pesticides and irrigation, are likely to lead to a pollution crisis, and they also clash with existing social relationships. Increased yields have led to riots in India, because the landlords refuse to pay higher wages for the extra work involved.

Whether or not man solves the pollution problem is not merely a question of scientific knowledge, but also of political understanding.

For this reason we must examine, if only briefly, the social and economic roots of society's production process and the major stages in its development.

Consciousness and Work

Work, the aspect of life from which so many feel alienated today, is none other than man's basic struggle to survive. The complexities of modern society, the trivial nature of many people's work, blinds them to the fact that labour is basic to man's being and defines his relationship to nature and the creative role he plays within it. It is, as Karl Marx says, a process 'in which both man and nature participate and in which man of his own accord starts, regulates, and controls the material reactions between himself and Nature'.[2] In this book, I stress the extent to which man has failed to regulate 'the material relations between himself and nature'. The main reason for this failure, I believe, is that man's labour, his associated technology, is being distorted by obsolete social relations. We must therefore analyse the labour process and the social relations that man develops around it. Man is not the only creature capable of labour; it is also characteristic of ants, spiders, bees, beavers, birds and many other species. But the labour of these creatures is instinctive, whereas the labour of man is capable of being totally conscious. In man, consciousness, language, social organisation and labour are all interrelated and capable of producing if necessary completely new behaviour patterns. The non-teleological process of natural selection has produced in man its opposite—a force which can consciously impress itself upon the natural forces which created it. Pollution is a result of this conflict of opposites, which will only be resolved either by the elimination of man, or by the raising of the human labour process to a greater level of consciousness.

We can distinguish two kinds of history—natural or biological history, and human history. Animals have a historical record of biological evolution. They have not changed themselves consciously, there has been no working out of 'inner needs', no seeking for any evolutionary goal. Man is developing out of this natural history.

We can measure man's degree of progress from the animal state by the degree to which he makes his own history consciously, in accordance with his desires, as opposed to being subject to blind nature. As Engels puts it, 'the more they (man) make their history themselves, consciously, the less becomes the influence of unforeseen effects and uncontrolled forces on this history, and the more accurately does the historical result correspond to the aim laid down in advance'. However, as Engels himself noted,

> 'If ... we apply this measure to human history, to that of even the most developed peoples of the present day, we find that there still exists here a colossal disproportion between the proposed aims and the results arrived at, that unforeseen effects predominate and that the uncontrolled forces are far more powerful than those set into motion according to plan. And this cannot be otherwise as long as the most essential historical activity of men, the one that has raised them from the animal to the human state and which forms the material foundation of all their other activities, namely the production of their requirements of life i.e. in our day social production, is above all subject to the interplay of unintended effects from uncontrolled forces and achieves its desired end only by way of exception, but much more frequently the exact opposite.'

Of course Engels was writing almost a hundred years ago (*c.* 1875-6). But how far have we really advanced? The thesis of this book is that whilst our technologies have advanced in power they have increased our capacity to produce unintended effects. Furthermore, the social advance necessary to control the new technologies has not been made. That in fact the socio-economic structure of our society produces social groups which consciously fight against the socially rational use of technology.

Engels explicitly related man's social weakness, i.e. society's failure to plan, to his own socialist views: 'Only conscious organisation of social production, in which production and distribution are carried out in a planned way, can lift mankind above the rest of the animal world as regards the social aspect, in the same way that production in general has done this for mankind in the specifically biological aspect. Historical evolution makes such an organisation daily more indispensable, but also with every day more possible. From it will date a new epoch of history, in which mankind itself, and with mankind all branches of its activity, and particularly natural science, will experience an advance that will put everything preceding it in the deepest shade.'

Marx's Theory of the Labour Process

The British classical economists of the eighteenth and early nineteenth centuries saw the relationship between nature and labour far more clearly than do the orthodox academic economists of today. They believed human labour to be the source of all wealth, yet at the same time they did not forget that the labour process begins with natural materials, such as soil and water. As Sir James Steuart, the eighteenth-century economist, put it: 'The earth's spontaneous productions being small in quantity, and quite independent of man, appear, as it were, to be furnished by Nature, in the same way as a small sum is given to a young man, in order to put him in a way of industry, and of making his fortune.'[3] This view, the labour theory of value, rejects the idea that capital is productive and stresses that the sole creative force in production is man.

As Marx and his followers took up and began to develop the labour theory of value so the academic economists began to drop it. In his study of John Maynard Keynes, Dudley Dillard writes, 'Only after the socialists took over the labour theory of value did it lose its respectability among academic economists.'[4] Without the labour theory of value it is impossible to analyse and understand the processes of production, and to discern the causes of the destructive aspects of our society. These destructive aspects therefore seem inexplicable, except in terms of 'selfishness' and 'stupidity'—which are again descriptions, not explanations. Environmental pollution must be understood by analysing its social roots in the process of production.

Let us take a brief look at Marx's analysis of the process of production. We have already said that human labour is unique in that it is conscious. Men labour to attain what they already possess in an idealised form in their imagination. Marx, in his analysis of the labour process, states that this process involves two other basic factors, besides purposeful conscious activity. These are the 'subject matter' of labour, and the 'instruments of labour'. By 'subject matter', Marx meant the naturally given subject matter of labour such as soil, water, and material which can be picked up, so to speak. Not all the subject matter of labour is in the natural state. As the productive powers of a society develop, so more and more of the subject matter has already undergone alteration by labour.

By 'instruments of labour', Marx meant: 'A thing, or a complex of things, which the labourer interposes between himself and the subject of his labour, and which serves as the conductor of his activity. He makes use of the mechanical, physical and chemical properties of some substances in order to make other substances subservient to his aims.' The concept of 'instruments of labour' goes beyond tools, including 'all such objects as are necessary for carrying on the labour process', e.g. domestic animals, workshops, roads, railways etc. The 'subject matter' and the 'instruments' make up the 'means of production'.

The production process is, to use Marx's words, the means by which 'Nature's material (is) adapted by a change of form to the wants of man'. The incorporation of human labour into the subject matter of labour produces something useful, a product which possesses a 'use value'. However, other 'use values', the products of earlier labour processes, generally enter into the present labour process as means of production. The use value which is the product of one labour process becomes the means of production in another. 'Products are therefore not only results, but also essential conditions of labour.'

Marx thus broke the labour process down into its most elementary factors, showing it to be 'human action with a view to producing use values, appropriation of natural substances to human requirements; it is the necessary condition for effecting exchange of matter between man and Nature; it is the everlasting nature-imposed condition of human existence, and therefore is independent of every social phase of that existence, or rather, common to every such phase'.

Marx developed in *Capital* his concept of the 'forces of production'. The 'forces of production' are not the same thing as the instruments (or means) of production, they are a dialectical concept embracing the total situation in which people equipped with particular production experience and habits of work operate the means of production. The concept 'encompasses *man* as a principal productive force'.[5] All too often, science and technology are confused with the productive forces, of which they are in fact merely an aspect. This leads to a failure to comprehend our contemporary situation, in which science and technology are rapidly advancing whilst society as a whole becomes increasingly unable to cope with a wide range of social problems such as poverty, under-development, and pollution. The fact is that the productive forces, as we shall explain in a later chapter, have effectively ceased to develop, and are stagnating.

Primitive Production

Benjamin Franklin defined man as a 'tool-making animal'; Gordon Childe points out: 'Man's emergence on the earth is indicated to the archaeologist by the tools he made. Man needs tools to supplement the deficiencies of his physiological equipment for securing food and shelter ... the first tools would presumably be bits of wood, bone or stone, very slightly sharpened or accommodated to the hand by breaking or chipping.'[6]

Marx, and later Childe, considered that fossil tools, the 'relics of by-gone instruments of labour', are extremely important in the study of vanished socio-economic formations. Primitive peoples can be classified according to the material from which they made their chief tools; bone, stone, bronze, and iron.

The discovery of fire was palaeolithic man's greatest technical leap forward. Fire enabled him to extend his geographical range, to explore caves for shelter, scare wild beasts, cook and use the night-time. Man had also taken his first step on the road to air and chemical pollution. If one can judge by the vast number of myths attached to it, (James Frazer recorded nearly 200), the discovery of fire and fear of its loss left its mark deep in the human psyche in a way no other discovery has since.

Men satisfy their basic needs by labour. Malinowski listed these most elementary needs: food, drink, rest, protection against excesses of heat, cold and damp, ensuring the reproduction of the race, and exercise of the muscles. These needs are satisfied *socially* by activity resulting from interactions between members of human groups.[7]

The most primitive peoples fed themselves by gathering wild

fruit, and elementary forms of hunting and fishing. Such peoples had only the simplest of tools and were therefore forced to take nature very much as they found it. Since they had no permanent food surplus, they were at any given moment only a few days away from starvation if they ran into a period of bad luck.

Richard Lee's study of the !Kung bushmen gives some valuable insights into the ecological basis of a primitive hunting and gathering economy.[8] Within a human group there is a division of labour and economic interdependence; the extent of such differentiation is a measure of the degree of advancement of a social organisation. Economics bases itself on a study of such transactions and exchanges within society.

Lee believes that a pooling of resources and breaking down of 'individual animal self-sufficiency' is unique to human society. It is, however, possible that its primitive beginnings could be traced in existing social primates. This is not, of course, to suggest that there is a linear relationship between human society and the social behaviour and organisation of any existing primates. In non-human primates every individual, except for the very young, forages daily for its own food. Even within social species such as baboons, which maintain themselves as a troop, each individual member is concerned only with feeding himself. As Lee puts it, 'the work rhythm is such that every individual must do some subsistence work on every single day of his adult life.'

Mankind has a radically different work rate which has evolved beyond the basic non-human primate condition. 'All human societies allocate some days to work and others to leisure, and in all human societies some people work harder than others... The sharing of food is part of a cluster of basic human institutions which also include the division of subsistence labour, the home base, the primary carrying device (for transporting foods to the home base for distribution), and the prolonged support of non-productive young and old people. These developments represent a *quantum step* [my italics] in human affairs, for their presence means that not *all* of the people have to work *all* of the time.'

The social division of labour opened up new social and economic possibilities. But a major impediment to the development of the social division of labour, beyond a very elementary form in primitive societies, was the inability to go beyond a subsistence economy and produce a surplus of food. In the next chapter we will examine some of the major steps in this process.

The relationship between labour (energy) and social structure has become extremely complex and hidden from view in modern society. The uncovering of the true nature of the exchange relations by scientists such as Marx has been greeted, not unnaturally, by the enmity of those who benefited by the existing social mystification. For this reason it is salutary to examine the basic human pattern of exchange in extant primitive societies.

According to Lee an economy

'... Exhibits an elementary form when the relation between the production and consumption of food is immediate in space and time. Such an economy would have the following properties: minimal surplus accumulation; minimal production of capital goods; an absence of agricultural and domestic animals; continuous food-getting activities by all able-bodied persons throughout the year; and self-sufficiency in foodstuffs and generalised reciprocity between local groups.'

Lee believes that although no contemporary society shows all these properties the !Kung Bushmen of the Dobe area of Botswana are a close approximation. The !Kung are hunters and gatherers and have persisted in such an elementary social form because of the extreme isolation and barrenness of their environment. They do not accumulate food surpluses, for 'food is almost always consumed within the boundaries of the local group and within forty-eight hours of its collection.'

For the !Kung food-getting itself is not a co-operative activity, but they co-operate in the consumption of food. The families pool the day's production and everybody including visitors receives equal shares of the available food. 'There is a constant flow of nuts, berries, roots and melons from one family fireplace to another until each person resident has received an equitable portion.' Lee found that, despite their primitive production process, the !Kung did not live on the brink of starvation; though they were of course extremely vulnerable to any natural disaster, since they had no surplus food stored for any such eventuality.

The !Kung apparently did not have to work particularly hard to produce their food. Lee made a careful study of their work effort, finding that 35% of the population was either too old or too young to work, and that the working members of the population produced the necessary food by working only 36% of the available working time.

Lee further observed that

> 'The group as a whole distributes the collective resources in such a way that the caloric needs of each age-sex class are met. In input-output terms this is a way of restating the classic dictum: *from each according to his means and to each according to his needs*. In principle the Bushman camp is a communistic society. In practice, sharing is never complete, but conflicting parties have the option of rearranging themselves spatially so that, when sharing breaks down, new groups can be constituted to ensure parity of production and consumption.'

What Lee showed was that for a modest work effort the !Kung Bushmen obtain an adequate living. Other research on hunting and gathering societies shows that the !Kung were not unique in this. In looking at such societies it is too easy to equate their lack of material wealth, i.e. private property, with wretched poverty. But such people can maintain health and happiness with a 'rudimentary technology'. Further, they have more leisure time per head than many industrial societies. The !Kung Bushmen, says Lee, spend 4-5 days a week in non-productive pursuits such as resting in camp or visiting other camps. They were a society at balance with themselves and their environment. Their morality objected to saving, since it was thought to be 'stingy or hard-hearted'. Sharing was the norm which kept everybody 'equally poor' and maintained their particular economic equilibrium. To break primitive communism the norm of 'all-for-one' and 'one-for-all' had to be suppressed and at the same time a technical breakthrough made in food production which would enable a consistent surplus to be produced. This breakthrough, the 'Neolithic Revolution', occurred several times independently from 15000 B.C. onwards and sounded the death-knell for the 'Eden' of Palaeolithic society. Man's 'childhood' was over and he entered 'adolescence', in which there was an increasingly great disparity in the distribution of the social product and increasing damage to the environment. The primitive communistic society remained only as a myth in the later class societies. Whether man can possibly outgrow his 'adolescence' and survive to develop in 'maturity' the balanced practice of his primitive communistic phase is the question which underlies my examination of the pollution problem.

4
From Hunting and Gathering to Industrial Production

The American ethnographer Louis Henry Morgan (1818-81) classified human history into three main periods; savagery, barbarism and civilisation. We have seen in the case of the !Kung Bushmen how savage society, a hunting and gathering economy, is capable of maintaining people in health, though at a low level of wealth and in small numbers. This form of society was the norm for man over most of man's million years of existence. No doubt peoples and tribes disappeared if and when they met with some unusual ecological catastrophe, disease, or climatic or other accident. However, man appeared to maintain some kind of balance with his environment and his impact on nature was puny in comparison with his recent 'achievements'. Then, in a period of less than fifteen thousand years, the productive forces of mankind progressed at an ever-increasing rate. At first sight it is hard to see how a creature which made such slow initial progress could attain such rates of growth.

The Importance of a Social Surplus

The economy of savages could only make very slow progress because it could not produce any social surplus of food, that is, they produced only enough food to satisfy their immediate needs. In fact, to have done so would put the precarious balance between their population and that of their food species at risk and lead to starvation. In the absence of such a social surplus only the most elementary division of labour, between the sexes—men hunting, and women gathering—could develop. As a result technical advances in the crafts were necessarily slow and painful.

However, since men did eventually go beyond savagery, there must have been a very slow accumulation of inventions, discoveries and knowledge, under rare favourable conditions, which led to increased food production and reduced physical effort, the first signs of an increased productivity of labour. One such invention would be the bow and arrow. Under these conditions a semi-settled, and later a settled, life could develop, which would allow people to accumulate tools over and above the limited amounts which could be carried by a migratory community. Thus a permanent social surplus could gradually have appeared. Such a surplus would make it possible to build up food reserves against famine, incidentally raising a new need to develop food storage and preservation techniques. Furthermore, an advance in the division of labour would then become possible. Some men could spend more of their time developing special skills, e.g. making arrowheads. Initially the surplus would not be big enough to allow full-time craftsmen. Under such conditions there developed craftsmen who wandered from community to community because none was rich enough to support them full-time. The first trade in the form of simple exchange by bartering would also start.

Such developments must have occurred many times, in many places, resulting in a more rapid increase in population. One can say that the density of population in a place is a good indicator of economic and social progress. Population growth and increased specialisation of labour increase the productive forces available to mankind. But an essential condition for such an advance is the appearance of a social surplus.

The Development of Agriculture

This modest increase in the social surplus was the *material basis* for mankind's greatest single technical advance until our own times, the development of agriculture. This has been called the 'Neolithic Revolution'; as a 'revolution' it was not a sudden violent change but rather 'the culmination of a progressive change in the economic structure and social organisation of communities that caused, or was accompanied by a dramatic increase in the population affected'.[1] In our discussion of the progressive rise in man's productive forces several such 'revolutions', the metallurgical, the urban, and the industrial, will be analysed.

Men discovered how to grow crops about 15000 B.C., and how to raise domestic animals about 10000 B.C. It is hardly necessary to point out that these developments took place spasmodically and independently in many parts of the world over a very long period. To make such discoveries presupposes that the discoverers already had a social surplus, since peoples on the brink of famine find it hard to devote resources to a future event—the harvest. Initially agriculture would have been a secondary activity, dependence upon hunting and gathering being maintained, and even when agriculture became the primary activity these would often remain as supplementary food resources.

The importance of the Neolithic Revolution lies in the fact that it reduced man's reliance on the caprices of nature, moving mankind from a 'passive' existence to one leading in the direction of increasing control, which would be amplified by a stimulation of crafts through further developments in the division of labour. There seems little doubt that in early agricultural communities the original communalism was carried on in the form of common ownership of the land. However, the Neolithic Revolution also gave a *material basis* for the development of social classes. The division of labour itself led to horizontal social stratification and eventually vertical social strata appeared on the scene. Furthermore, the social surplus developed by agriculture allowed mankind to progress to make further major social and technical innovations such as metallurgy, trading and towns.

Exactly what the circumstances of the discovery of metallurgy were still remains shrouded in the mists of time. The speculations of Mandel sound plausible—that with the development of agriculture, 'Men released from the degrading servitude of hunger, were able to develop their innate qualities of curiosity and technical experimentation. They had long since learnt that it was possible to cook certain kinds of clay in the fire to make pots. By subjecting different kinds of stones to the fire they discovered metals, and then their wonderful capacity for being made into tools.'[2] The discovery of the various metals and how to work them occurred spasmodically over thousands of years. Copper was discovered in the sixth millennium B.C. in the valleys of the Tigris and Euphrates and in the Nile Valley, tin and bronze (a copper-tin alloy) in the third millennium B.C. in Egypt, Mesopotamia, Iran and India and finally iron (c. 1300 B.C.) among the Hittites and also among races in the Black Sea region.

The use of metals exerted a profound social effect, particularly in agriculture, crafts and warfare. In agriculture, the development of ploughs with metal shares greatly increased the range of soils that could be worked. The development of the iron plough in the seventh and eighth centuries A.D. in Europe made agriculture possible in regions with heavy soils. A great range of crafts developed around the use of metal.

The Rise of the City

The next important steps made by man have been termed the 'Urban Revolution', the process by which pre-literate agriculturalists living in villages came to form larger and more complex communities—'cities', which formed the basis of civilised society. The other criteria of a civilised society are: classes of full-time specialists, e.g. kings, priests, craftsmen, warriors, etc., who are exempt from the ordinary subsistence tasks: social mechanisms for concentrating the 'social surplus' in the hands of an élite by means of taxes and tributes; monumental public buildings for the kings and gods: the elaboration of exact predictive sciences and writing; extensive and regular foreign trade; the emergence of political organisation based on residence rather than kinship.

The development of the 'urban revolution' was uneven; we find 'cities' developing independently at a number of places over a period of several thousand years. The *pioneering* developments were based upon certain local peculiarities. They tended to be in fertile, irrigated river valleys, like the Tigris and Euphrates valleys.

A complex of factors was responsible for the drastic political changes of the urban revolution, e.g. trading, warfare, exposure to natural hazards such as droughts, floods, insect-plagues, a relatively dry climate and grains permitting extensive storage of foods, a shortage of water and easily irrigated land which formed a local source of economic inequality giving an advantage to certain groups.

The Pre-Industrial City

According to Lewis Mumford the earliest cities were centres of royal and priestly powers and were initially control centres rather than marketing or manufacturing centres. At their heart was a walled 'citadel' which contained a temple, a palace and the granary. The centre of political, religious, scientific, military and economic power resided in the 'citadel'. Thus the ruling class protected itself from those it exploited and from outside invaders. However, as the cities developed they redeveloped structures found in the villages. The eating, drinking, sleeping and sexual functions of the village household developed specialised city forms; café, hotel and brothel. The city developed to a higher level features such as the market, craft workshops, and exchange media such as money. The city became the 'container and transmitter of culture ... increasing the variety, the velocity, the extent and continuity of human intercourse'.[3]

With the city came the first pollution problems. The sewage wastes of the villages are generally taken care of by the need to fertilise the land and by the self-purification processes of the rivers. But the relatively large number of people living at close quarters in towns creates new problems of hygiene and sanitation. The city becomes a prey to contagious diseases, a problem solved only within the last hundred years. It was thus necessary to arrange at least a minimum programme of systematic disposal of garbage, excrement, rubbish and the dead.

As early as the sixth century B.C. Rome had an extensive system of sewers which flowed via a giant drain over a thousand feet long, ten feet high and twenty feet wide, the Cloaca Maxima, into the Tiber. Citizens wishing to make a connection between their property and the city drain had to pay a special tax—the *cloacarium*. When cities developed in temperate climates houses needed heating in winter and air pollution began to be noticed. In London the first air pollution regulations against the burning of coal occurred in the thirteenth century.

The citadel often contained piped water, baths, water closets and private toilets. It was over a thousand years before such facilities became available to the 'masses'. As we shall see later, the English industrial towns, for example, grew up without adequate drinking water or sanitation, and when they began to be cleaned up the problem of pollution was all too often merely shifted to the rural and aquatic environment. In addition to their political, religious and educational functions pre-industrial cities were marketing and handicraft centres. Some were specialised centres of religion, politics or education. In pre-industrial societies the urban population rarely reached 10% of the total.[4] This is because the amount of surplus food production was limited by unmechanised agriculture, transport and food preservation and storage problems.

The bulk of the energy used in production, within pre-industrial cities, was animal energy. Where wind or water power was available, it was probably used outside the cities. Production within the pre-industrial cities took the form of petty-commodity production, that is, it was dominated by craftsmen working in their own homes or nearby shops, within the regulations and limits of a craft guild. The guilds controlled conditions of work, and methods of production and sale. A craftsman usually made a whole product, with the help of apprentices, and was his own master. Division of labour only occurred between crafts, not within a craft. Relatively few middlemen could be supported. Guild membership or an apprenticeship were prerequisites to the practice of any occupation, thus allowing the development of a monopoly of particular craft skills.

We have seen that the surplus product of agriculture was the material basis of all civilisations. In theory this surplus product could take the form of labour services, payment in kind or money. However, since the bulk of the population lived under a natural economy, from what they produced themselves, rent was paid to the landlord either by payments in kind or by unpaid labour.

The lord was able to use up this surplus for his own consumption or sell it on the market. Only such people were in a position to buy the products of craftsmen working in the towns. Consequently much of the production was luxury goods, and the lack of a mass market limited the development of craft production to petty-commodity production. As a result the civilisations of Ancient Greece, Rome, Byzantium, Islam, India and China were still basically agricultural despite their splendour.

Mandel says that 'so long as agricultural surplus product retains the form of goods in kind, trade, money and capital could only develop superficially'. As a result history witnessed a succession of civilisations which rose and fell, each leaving and adding some increment for succeeding societies. This meant that the advances in the forces of production were 'gradual', compared with what Western Europe has experienced since the fifteenth century. In Western Europe there developed a new form of socio-economic formation, capitalist society. Capitalist society was the result of a developmental process in which the dominance of petty-commodity production by the craft guilds was broken and replaced by the commodity production of a manufacturing industry.

In all previous socio-economic systems the manufacturer was primarily concerned with a commodity's use value. Under capitalism the use value is incidental as far as the primary driving forces of the system are concerned. The capitalist concerns himself primarily with the exchange value. The capitalist starts with a sum of money which is invested in production of a commodity which is sold for *a higher value of money*. He fails if he does not increase his capital in the process of production. This is the essence of commodity production. Historically the development of capitalism involved the *dominance* of commodity production and the creation of a working class which possessed only a single commodity for sale—its labour power.

The Development of Capitalist Society

The fact that Western Europe provided the right conditions for capitalist society to develop is of some interest, because it illustrates very vividly the uneven manner of development of the productive forces of mankind.

Other civilisations developed usurers' capital and merchant capital, domestic industry, wind- and water-driven machines—most of the ingredients found in Europe prior to its 'take-off'. The missing factor found in Europe was that the natural economy of the peasant populations began to be broken down by the penetration of a money economy. Europe was the only area where agricultural surplus assumed for any period the form of a money rent.[5]

This development of money rent alone is not enough to force the development of commodity production. But it improves the possibility of the concentration of wealth in the form of money, capital, by a new possessing class, the 'bourgeoisie'—a vital development in social evolution.

Outside Europe, since the bulk of the population never got money and therefore did not participate in the market, trade and commodity production never got beyond their major concern with luxuries. There was no possibility of a mass market developing and with such a limited market capital often found land or usury a more profitable investment than production.

In some societies the easy availability of slaves or masses of cheap labour meant that there was no need for other than a sporadic development of machinery. A further factor in Europe's favour was that the feudal society, unlike the Roman and Asiatic civilisations, did not have a strong central government capable of wielding despotic powers against their bourgeoisie.[6] In fact, in their struggles to develop central governments the kings of Europe relied very heavily on their bourgeoisie for financial aid.

In Europe there occurred a unique conjuncture of events, and the capital, the market, the technology and the labour forces developed within the same historical process. By the sixteenth century, the advance of the bourgeoisie in several Western European countries was virtually unstoppable, and their economic victory was, where necessary, backed up by political revolution. The expansion of European-based trade, from the eleventh century onwards, speeded up the development of a money economy in Western Europe. Associated with this expansion were increased requirements of precious metals, the opening up of new trade routes, the discovery of the Americas, and the plundering of Latin American and Far Eastern civilisations. Commercial capitalism began to develop on the basis of the vastly extended world market. This development was not a simple, smooth one; at times massive inflation and the ruination of nobles, merchants and craftsmen ensued.

In England, from the fifteenth century onwards, the break-up of feudalism, changes in agricultural practice, and the downfall of the craftsmen brought into being, over a two-hundred-year period, a proletariat. The proletariat were propertyless, capable only of begging, crime, and selling labour power to owners of capital.[7] Analogous developments took place earlier or later in other Western European countries.

Mining was the area where the commercial bourgeoisie had its greatest early success in taking possession of the means of production. By the seventeenth century factory manufacturing was developing alongside the domestic industry. Workers who worked with tools and raw materials provided for them became wage-earners, losing their status of independent craftsmen. In petty-commodity production there is *social* division between different crafts but little division of labour within a craft. Manufacture, however, encouraged the sub-division of crafts into a series of labour operations. This increased production and further reduced its costs by substituting for expensive skilled craftsmen a less well-trained but more specialised labour force. At first, however, manual labour was still the predominant force in manufacturing.

The technical improvements accumulated in Europe since the fifteenth century, particularly in the use of water power, were only applied sporadically whilst social and economic conditions did not favour a large-scale flow of capital into industrial production.

Capital entered the sphere of industrial production as the

bourgeoisie realised that they were no longer confronted by a stable market but by an apparently ever-expanding one. Initially the most rapid area of capital investment and consequently of technical innovation was in mining and metallurgy. The steam engine and the railways were developed initially for the mining industry. In Britain an ever-expanding coal and iron industry was followed, by the end of the eighteenth century, by the mechanisation of the textile industry, its power being provided initially by water and then by steam. With the application of steam power and the development of the factory the rise of the industrial society became possible. What had been small unknown villages became almost overnight the largest cities of the land and the centres of the new transport system of railways.

Industrial capitalism did not develop evenly everywhere at the same time and to the same extent. By 1850 the industrial revolution had entered France and Belgium. By 1900 it had reached Germany, Sweden, the USA, Northern Italy, Russia and Japan. Today very few countries remain untouched. Those countries which pioneered were overtaken. The dominant economy today, the United States of America, was virtually virgin territory inhabited mainly by 'savages' when Holland and England were developing commercial capitalism in the sixteenth and seventeenth centuries.

Capitalism and the Rise of Rationality

The rise of capitalist society was accompanied by a general rise in rationality which began to break the stifling grip of traditionalism. Werner Sombart described the traditional mode of thinking as being one concerned with perpetuation rather than change. 'In deciding on some undertaking or activity, a man does not look in front of him, to his goal, he does not exclusively consider the purpose of his decision, but he looks back to examples and experiences of the past.' The transition from a natural economy to a money economy enabled a former multiplicity of human goals to be replaced by a single aim—that of making money. A more analytical approach could thus be used when evaluating the usefulness of any activity. There arose a growing tendency to evaluate such activity, not according to tradition, but according to whether or not it would yield more money than was invested in it.

The concept of profit enables different activities to be compared by a uniform quantitatively measurable end. Sombart believed that the development of such business calculation was a vitally important factor in the development of modern science; the 'constant application of a conception which treats all phenomena as quantities, quantification, an idea which has brought to light all marvels of nature and which here, for the first time in history, has become quite clearly the basic idea of a particular system'. It is quite conceivable that without the prior development of double-entry book-keeping the rise of modern science would have been greatly retarded. There are, of course, a great many other ways in which science has been stimulated by the development of capitalism, for example the provision of research funds and new problems.

The development of capitalist society encouraged the intensification of economic rationality, together with the parallel and intimately related development of science and technology, which led to an unprecedented increase in the productive forces of society. But the fatal contradiction of capitalist rationality was that it was confined to the private interests of individual enterprises, which predominated over wider social interests. Thus it was not surprising that the changes brought about by the new order, and by industrialism in particular, were often against the public interest.

PART TWO

The Current Situation

5
The Pollution Crisis in the Cities

With the birth of the city, man created a unique environment for himself. His relationship with nature, his place in the myriad biological linkages between the world's manifold life-forms, became buried beneath artifacts and manners. The city became the centre of culture and civilisation, the storehouse of human experience in the arts and the sciences. It became the centre of governments and workshops and eventually of large-scale industry. But the belief that man can cut himself off from nature is fallacious. He can no more do this than his mind can soar away in independence from his body.

The human population, whether it be living in a hunting and gathering community in the Amazonian tropical forest or in a giant metropolis, is a population of animals. There are certain biological laws that we are subject to on both the population and individual levels.

Many aspects of the so-called urban crisis are a reflection within society of these laws. The first law is that all life-forms are organised in ecosystems. The city, the urban environment, is an ecosystem, but one in which the range of life-forms is strictly limited, with a high preponderance of humans. This puts the system in a high state of unbalance, for without massive injections of energy and material from outside it could not be maintained. However, besides taking in food and other supplies, biological systems must be able to get rid of their waste products if they are to survive. Under the conditions of urbanisation, even natural wastes, such as faeces, urine and uneaten food, are accumulated in quantities too large to be dealt with by the natural biological degradation processes. On top of such natural wastes, urban man is also continually adding an ever-increasing range of synthetic wastes.

It has been calculated that an American city with a population of one million 'feeds' on: 625,000 tons of water, 2,000 tons of food and 9,500 tons of fuel (comprising 3,000 tons of coal, 2,800 tons of oil, 2,300 tons of natural gas and 1,000 tons of motor fuel). Without all this the city would grind to a halt. After 'digesting' this daily feast the city excretes: 500,000 tons of sewage (i.e. water polluted with 120 tons of suspended solids), 2,000 tons of refuse and 950 tons of air pollutants (composed of 150 tons of particles, 150 tons of sulphur dioxide, 100 tons of nitrogen oxides, 100 tons of hydrocarbons, and 450 tons of carbon monoxide).[1]

Besides such biological needs there are social and economic requirements which correspond to the mode of production. The resultant interaction between human needs on the biological and on the social levels is not always harmonious. They often conflict. Urban pollution is a symptom of this conflict.

The Growing Urban Sprawl

The concentration of people into ever-growing urban areas is causing the basic biological cycling systems and man-made services to break down under the ever-increasing size of the job they are being asked to do.

All parts of the world seem to be experiencing a continuing

migration from the countryside to the cities. For example, a recent Council of Europe report[2] stated that in France 150,000 people a year are leaving the land for the city, and that by 1985 80% of the French population will live in towns. At the same time deteriorating conditions in the heart of big cities lead to people moving out to new suburbs, and thus the urban areas eat into the countryside. In the Netherlands, half the population live in one region, Randstad Holland, which occupies only two-ninths of the country's area. Built-up areas take up 9% of the total land area and it is estimated that by the end of the century 25% will have been built upon if present trends continue. Holland is by no means unique.

Giant urban sprawls—like London and South-East England, South-East Lancashire, New York and Los Angeles—are developing in most countries in the world. The term megalopolis has been applied to them, and if current trends are allowed to continue we will create Ecumenopolis, a single world-wide city.[3]

A question increasingly asked today is, are these urban sprawls the most effective and healthy way of organising society? The advances in basic hygiene in the nineteenth century, and in chemotherapy in this century, have led to a great reduction in infant mortality but it appears that there is an increase in neoplastic diseases. Professor I. R. Passino has said 'a city is a large disease factory'.[4] Chronic lung diseases, probably the greatest health problem in the industrial regions of Northern Europe, are most common in urban areas. There is also a marked increase in nervous diseases in urban regions. According to Professor Jean-Pierre Ribant, between 10.9% and 25% of the inhabitants of large urban centres suffer from more or less serious mental disturbances. Thousands of town-dwellers are killed or maimed in accidents in the transport systems, at work and in the home. It is also claimed that industrial society—the acme of human progress—has resulted in an increase in the number of people who are overweight, disabled, sick and mentally deficient. Passino proclaims that 'We are engaged in a form of chemical and biological warfare against ourselves, and especially against our children'.

Air Pollution

Although the atmosphere has apparently changed and evolved greatly over time, in historical time its basic overall composition has altered only within rather narrow limits. However, as we shall see later, even small changes can have very profound climatic implications. The atmosphere is a dynamic system which can take up many substances, particles, liquids and gases, from natural and man-made sources. Within the air these substances can react together physically and chemically; they may remain in the air for varying lengths of time and may leave the atmospheric system, falling into the sea or on to the land, or being absorbed by living creatures including man. Although in this chapter we are mainly concerned with the atmosphere above our cities, we must remember that the urban air masses connect up with and form part of regional and global air systems. Materials taken up by the atmosphere above the cities may be carried elsewhere and may even be affecting global weather systems.

Air is a mixture; dry clean air contains sixteen gaseous components. But only two of them account for almost 99% of the composition of air—nitrogen (78.09%) and oxygen (20.94%).[5]

The proportion of nitrogen and oxygen is not likely to be changed in the foreseeable future by man's actions but relatively significant changes could be made amongst some of the other components of air.

There are no world figures of the total amounts of pollutants poured into the atmosphere by man. The United States Public Health Service reports that in the USA the most common primary pollutants (by weight) are carbon monoxide, sulphur oxides, hydrocarbons, nitrogen oxides and particles. Motor cars, industry, electrical power stations, space heating and refuse disposal are the major sources of air pollution in the USA and other advanced countries.

Since most of the major air-polluting activities are more concentrated in urban areas than rural it is not surprising that city air is more contaminated than country air. In the US the average concentration of particles in urban air is approximately three times as great as in country air.

The Biological Function of Air

To understand the effects of atmospheric pollution on man a little elementary biology is necessary. We have seen that about a fifth of the air consists of the gas, oxygen. Nearly all living things, and all higher life, require oxygen for the liberation of the energy necessary to keep them alive. The gas carbon dioxide is a byproduct of this respiration process, and has to be eliminated from the organism. In order to survive, each individual cell must respire, i.e. be provided with the means to receive food and oxygen and get rid of carbon dioxide. In man this is accomplished by our breathing and circulating systems—the breathing system collects and gets rid of gases from and to the outside world, and our circulating system transports the gases to and from the internal cells of the body.

The breathing system basically consists of our two lungs and the air passages leading to them. The air passages begin at the nose and mouth, branching many times into very fine tubes (bronchioles), which end in masses of minute air-sacs (alveoli). The purpose of this structure is to pack a very large surface area, about 100 square yards, into the small volume of the chest cavity. Each alveolus is surrounded by a dense network of blood-vessels. It is here that gases are exchanged between the air and blood. As much blood as possible is brought into close proximity with the air in the alveoli. The average man breathes 22,000 times a day, taking in 35 lb of air. The red blood-cells are able to carry oxygen by combining it with a chemical, haemoglobin. Haemoglobin links up with oxygen in the lungs and is able to release the oxygen to the body cells. Carbon dioxide is carried away from the cells in the blood and is released into the lungs and breathed out.

If this system can get oxygen rapidly from the air to the cells of our tissues it may also under certain circumstances be the avenue for poisons. Thus the breathing system is endowed with a series of complex protective mechanisms.

Our nose and upper respiratory tract are able to trap large particles. Small particles, less than three microns in diameter, which reach the deeper parts of the lungs can be dealt with in a number of ways. They can, for example, be trapped in mucus and carried away from the lungs by the beating of microscopic hairs, cilia, which line the breathing tubes.

Such mechanisms proved to be adequate over the greater part of man's history. However, the onset of atmospheric pollution posed these mechanisms with problems which they could only partially cope with.

The Effects of Pollution on Health

The concept of toxicity is an extremely difficult one. Only rarely in pollution-induced toxicity are the causal connections simple, mechanical and clear-cut. The clear-cut cases usually take an acute form, for example, a stove burning coke gives off carbon monoxide fumes which poison people in badly ventilated rooms. The coroner, after hearing the medical evidence in such cases, concludes that they have been killed by carbon monoxide. But how is the legal mind to cope with the fact that John Smith of Middlesbrough, a much polluted city, dies at the age of sixty-three of chronic bronchitis? The causation behind John Smith's untimely death is a complex of many factors: a lifetime of hard physical labour in a factory, bad housing, poor nutrition in childhood, as well as constant exposure to Middlesbrough's bad air. This complexity is often used as a pseudo-scientific cloak behind which governments, industrialists, and others shelter in their efforts to avoid legislation which would inhibit their 'right' to pollute.

The bad effects of smokey air have been known for centuries. The kings of England fought a losing battle to keep the citizens of London from burning coal for some time in the fourteenth and fifteenth centuries. However, it was not until the work of Russell in the 1920s that this knowledge began to be put on a scientific footing. He was the first to show that cold foggy weather in British cities was associated with a rise in the number of people dying from respiratory diseases and other causes.[6] Since that time there have been a number of terrible disasters which helped provide in some cases the necessary data for firmer legislative action to control air pollution. The most infamous cases happened in the Meuse Valley, at Donora and Poza Rica and in the great London fog of 1952.

The Great Fog Disasters[7]

The Meuse Valley is a narrow winding valley in Belgium. It is surrounded by low hills and its industries include coke-ovens, blast furnaces, steel mills, glass factories, zinc smelters and sulphuric acid plants. On December 1st, 1930, there was unusually calm weather and a thermal inversion (normally the air temperature decreases with altitude; a thermal inversion is when temperature increases with altitude). Such weather conditions can lead to a stationary mass of air in which air pollutants build up and accumulate. On this occasion the condition lasted several days. Within a week sixty people had died of respiratory illnesses. Many more were made ill, suffering chest pains, coughing, shortness of breath and irritation of their eyes and noses. Exactly what the killer substance was nobody seems sure; possibly it was a combination of several pollutants. Some think that sulphur dioxide was the chief culprit; dissolved in fine droplets of water and with fine smoke particles it can give a mist of highly corrosive sulphuric acid, which may penetrate deeply into the lungs. The events and implications of the Meuse Valley were noted by a few people. J. Firket wrote 'The public services of London might be faced with the responsibility of 3,200 sudden deaths if such a phenomenon occurred here.'

Donora, Pennsylvania, in the United States was the scene of the next killer fog. This was on October 23rd, 1948. Again the location was a steep-sided valley with windless conditions, a thermal inversion and a fog. After two days, says Breton Roueche, the fog had turned into 'a motionless clot of smoke'. Almost half the population became sick, mostly complaining of coughs, of chest, throat, nose and eye irritation, of vomiting and nausea. By the end of the third day twenty of them had died.

In Poza Rica in Mexico, on November 21st, 1950, hydrogen sulphide escaped from a chemical plant around 5.0 a.m. It was foggy and there was little wind and the gas concentrated around the surrounding houses where many people were still asleep. Twenty-two people died and a further 320 had to go to hospital for treatment.

The greatest air pollution disaster of all time occurred in London between December 5th and 9th, 1952, precisely as predicted by Firket twenty years earlier. A combination of windlessness and temperature inversion gave rise to a very thick fog. During this period there was an increase in admissions to the hospitals for heart and respiratory illness, chiefly of the elderly. When the mortality figures for the period of the fog were examined it was found that there had been 3,500-4,000 more deaths than usual for the time of year. Of course, London was 'famous' for its fogs and researchers found that there had been similar happenings before, e.g. 300 'excess' deaths after a fog in 1948. Several more killer fogs were to come to London, though never on such a scale, the last being in 1962 when there were 400-700 'excess' deaths. Since then the implimentation of the Clean Air Act has eliminated the killer—London fog.

Between November 27th and December 10th, 1962, air pollution swept across the globe. It first affected Washington, Philadelphia, New York and Cincinnati in late November. On December 2nd to 7th Rotterdam suffered thick persistent fog; between December 3rd and 7th fog came to Hamburg, Paris, the Ruhr, Frankfurt, Prague and London (where many died, see previous paragraph). The fog caused illnesses in all the places mentioned. The exact cause and full global extent of these events are not known; the fog did not affect Australia and the American West coast. In the summer of 1970 there was a similar global series of smogs, which affected New York, Milan in Italy, Japan and Sydney, Australia.

Bronchitis and Air Pollution

Bronchitis is so common in industrial Britain that it has been called 'the British disease'. It is an inflammation of the mucous membrane lining the air-tubes (the bronchi) leading from the windpipe to the lungs. The mucus secreted by the special cells lining the bronchi neutralises and dissolves gaseous pollutants and traps foreign particles. Chronic bronchitis is a persistent exaggerated condition of this natural defence-mechanism. The cilia which beat the mucus along to the throat, from where it can be spat out, are slowed down, causing a build-up of mucus laden with bacteria and pollutants. Bronchitis provides conditions for the development of potentially more dangerous ailments.

Chronic bronchitis has shortened many people's lives and greatly reduces their capacity to get the best out of life.

A British study found that the mortality from bronchitis was 82 per 100,000 in industrial zones as opposed to 46 per 100,000 in non-industrial zones.[8] In London a higher percentage of people die of bronchitis in the most polluted districts than in the cleaner areas.

It should be mentioned, however, that the British findings have not always been confirmed in other countries. Just why this should be so is not clear. It may be due to differences in recording the necessary medical statistical data required for such studies. It could be that air pollution causes bronchitis only under the damp climatic conditions of Britain. Evidence from the continent includes an Italian study which found that bronchitis was up to ten times more frequent in the industrial zones of Genoa than in the non-industrial areas. In Federal Germany the incidence of bronchitis amongst town children (3.53%) was greater than amongst country children (2.49%), whereas amongst adults it was almost the same in town (7.7%) and country (7%).

It would seem that there are many complicating factors in the causation of bronchitis. Bad housing conditions and nutrition which result from poverty and overcrowding certainly contribute, and probably make it harder for the body to withstand the rigours of polluted air in a British winter. Certainly bronchitis is less frequently a rich man's disease than a worker's.

Air Pollution and Lung Cancer

The industrial countries are experiencing an epidemic of lung cancer. The major factor in this increase is the increased consumption of cigarette tobacco. But some scientists think that lung cancer is also caused by air pollution, though to a far lesser extent than by smoking. Nobody is sure just what substances in the air are the agents responsible for causing cancer. The level of lung cancer has been said by Professor Petrilli to be highest where the levels of 3, 4-benzpyrene, beryllium, arsenic and molybdenum in the air are greatest. However, not all experts believe air pollution to be a contributory cause of lung cancer; whilst there is evidence for an 'urban factor' which contributes to lung cancer (for example, Waller found that the lung cancer mortality rate for Englishmen living in cities was about two times higher than the rate for men living in a rural environment), they believe there is no firm evidence that this factor is air pollution. It would seem that if you want to maximise your chances of not getting lung cancer, don't smoke or live in a town and don't live in regions with polluted air.

There may well prove to be some most unexpected sources of carcinogens which can cause lung cancer. In Mexico, for example, there are quite high rates of lung cancer amongst women and it is suggested that this is because they spend many hours each day over smoking cooking stoves. Fumes produced during cooking include carbon monoxide, sulphur oxides, sooty and oily aerosols. Some evidence from the United States seems to support the Mexican finding, for the lung cancer rate amongst American rural non-smoking women is higher than in the corresponding class of men. Normally men have a higher lung cancer rate.

Benzpyrene is a hydrocarbon, a chemical containing hydrogen and carbon; it is also a known carcinogen. It has been shown experimentally to produce cancer when injected into test animals. It is found in cigarette smoke, smoke from certain industrial processes and the exhaust gases of motor cars, and is something that we often breathe into our lungs. But it has proved extremely difficult to show experimentally that inhalation of this compound will produce tumours of the lung. Some Russian scientists have obtained experimental lung cancers by sending the carcinogen into the lungs in association with small particles, which cause chronic inflammation of the mucus-cleansing mechanisms of the lungs. This allows lumps or 'depots' of the carcinogen to form in the lungs.[9] Maybe this is what happens to smokers and city dwellers.

Only in the last twenty years or so have systematic investigations of air contamination by carcinogens been made by scientists. The amounts found are determined by local industrial or traffic conditions. In the Russian town of Makeevka, where an old coke plant created a large amount of air pollution, after the installation of pollution control equipment the level of 3, 4-benzpyrene in the neighbouring air was reduced 1,000 times. The average level of 3, 4-benzpyrene in the air of Soviet cities ranged from 4.43 micrograms per cubic metre ($\mu g/m^3$) in Leningrad to 0.12 $\mu g/m^3$ in Angarsk.

In a large industrial city several kilogrammes of 3, 4-benzpyrene can be discharged into the air each year. Much of it comes back to the ground, contaminating soil and water. In Moscow the soil is heavily contaminated by benzpyrene; levels of up to 346.5 $\mu g/kg$ of soil have been found there. The level in the oldest parts of the city is 2-3 times greater than in the new parts. In wild isolated places no benzpyrene was found in the soil, but in soil near some oil refineries the benzpyrene level was as high as 200,000 $\mu g/kg$.

Benzpyrene can settle from the air on to plants, and it can also be taken up from the soil by plants. It is not easily washed off, for after washing contaminated plants in hot water for 30 minutes much of their original benzpyrene content still remains. One worrying discovery is that although some soil micro-organisms break down benzpyrene into innocuous substances there are some which are able to accumulate it. It is extremely disturbing that such potentially dangerous chemicals are allowed to escape into the air, and that they eventually settle in the soil where they might be accumulated by soil bacteria and later taken up by vegetation, ultimately entering the food of domestic animals or even that of man. Shall we in time discover that lung cancer is not the only form of cancer to be associated with air pollution?

As we have seen, there is statistical evidence which shows that after correction for social and economic factors city dwellers suffer in greater measure from certain diseases than those who live in the country.[10] It seems reasonable to suggest that this is due in large measure to the unhealthy effects of continuously breathing air contaminated by polluting chemicals. The next section looks at what we know about the biological and medical effects of some of the major air pollutants.

The Biological Effects of Specific Air Pollutants:

Carbon Monoxide
Carbon monoxide is one of the most common pollutant gases.

Most of it is produced by the burning of substances containing carbon, e.g. coal and petrol. The average concentration in the air is between 0.1-0.2 parts per million (ppm). There is no known mechanism whereby carbon monoxide can be removed from the atmosphere. At the current rate of addition the atmospheric concentration of carbon monoxide could be increasing by 0.03 ppm every year.[11] Unfortunately present analytical techniques are not sensitive enough to check whether such an increase is in fact occurring. The amount produced by natural sources seems minute[12] and there has been a continuous general increase in man-made carbon monoxide produced over the last 100 years. In the United States the level is expected to double over the twenty-year period 1960-1980 to 250 million tons per year unless major changes are made in automobile exhaust emissions. We can certainly expect similar rates of increase in other parts of the world. A standard work on the subject claims, 'no other gaseous air pollutant with such a toxic potential as carbon monoxide exists at such high concentrations in urban atmospheres'.[13] Some scientists think we are already far closer to the danger level for carbon monoxide than we are for any other gaseous pollutant. Levels of 30 ppm carbon monoxide are often found in urban air, and this is close to the maximum allowable level for industrial exposure in the USA, 50 ppm over an eight-hour day.

In acute doses carbon monoxide is a killer; it is a favourite with suicides who run their car engines in a sealed garage. Symptoms of exposure to high concentrations are headache, dizziness, lassitude, flickering before the eyes, ringing in the ears, nausea, vomiting, palpitations, pressure on the chest, difficulty in breathing, apathy, muscular weakness, collapse, unconsciousness and death. Carbon monoxide kills because it interferes with the oxygen-carrying capacity of the blood. Haemoglobin will combine with carbon monoxide in preference to oxygen. The carboxyhaemoglobin (carbon monoxide plus haemoglobin) does not break down very quickly and thus the blood's capacity to carry oxygen diminishes. As a result the body cells do not get enough oxygen and the nasty characteristic symptoms develop.

If the effects of large doses of carbon monoxide seem clear enough the same cannot be said of the chronic effects of small doses over long periods. It is possible that current levels of carbon monoxide could already be high enough to injure people suffering from heart or respiratory illnesses. Current 'safety' levels often do not take account of the fact that a considerable percentage of the population is already sick.

Belgian workers exposed to high carbon monoxide levels at work often have to be retired four or five years before the normal age because they suffer from cardiovascular disease. It has been suggested that carbon monoxide might be partially responsible.[14]

Such studies are complicated by all sorts of factors; smoking a single cigarette can raise the carbon monoxide level in the blood.

Sulphur Dioxide

Sulphur dioxide is a nasty, pungent gas, which with water and air can form a powerful acid. This is one pollutant which recent anti-smoke laws have not always succeeded in removing from the air of our cities. During the 1960s the sulphur content of the air over most of Europe rose and the acidity of the rainfall increased. Britain seems to have been an exception and the average urban SO_2 level in Britain was reduced by about a third from 1959-69.

Not all the sulphur dioxide in the global atmosphere results from human activity. It is thought that about 80% of it is derived from another gas—hydrogen sulphide, the one that smells of bad eggs, which is produced naturally by decaying organic matter. Most of the remaining 20% is a pollutant produced by burning fuel which contains sulphur; a little (though it can be important locally) results from smelting non-ferrous metal and oil refining.

Normal clean dry air contains 0.0002 parts per million (ppm) of sulphur dioxide whereas in 1965 the average sulphur dioxide in Chicago's air was 0.130 ppm and in Washington's, 0.046 ppm. During the great London fog of 1952 a sulphur dioxide level of 1.34 ppm was reached.

Exactly what harm the sulphur dioxide does to man at these levels is not known; but it is surely better to play safe and get the levels reduced even if it means increased fuel charges.

Some industrial workers have been exposed to levels as high as 36 ppm for years. As a result of this they suffer from nasopharyngitis, coughs, expectoration, shortness of breath and other signs of respiratory disease. These levels are many times greater than the sulphur dioxide levels in city air.

Here we are again faced with the old toxicological problems of small amounts of a poison taken continuously over very long periods. People living in polluted city air do suffer more from certain respiratory diseases, but has the sulphur dioxide anything to do with this? Experiments have been done on animals and men so as to find out more about the effects of low levels of SO_2.

One finding is that animal species vary greatly in their susceptibility to sulphur dioxide. Rats are the most resistant, and guinea-pigs the least, amongst animals commonly used by experimenters. Man is said to be even more sensitive than the guinea-pig. In one experiment mice breathed in air containing 25 ppm of sulphur dioxide for 1,100 days with no ill effects. Russian toxicologists have pioneered methods of testing very small doses of suspected poisons on the nervous system by examining their effects on conditioned reflexes.[15] They found that 0.2 ppm is the lowest level that affects conditioned reflexes in the brain *cortex*. Breathing air containing 10 ppm sulphur dioxide for ten minutes actually increases the rate of breathing and pulse in man. There are two possible mechanisms by which sulphur dioxide might damage the lungs. One is that it could be converted into droplets of sulphuric acid which are inhaled. The other is that sulphur dioxide might be absorbed on to small particles. These particles could be drawn deep into the lungs where the sulphur dioxide might then build up into greater concentrations than are found in the air itself. Certainly the effect on buildings of continuous exposure to SO_2 suggests that sulphur dioxide could be a threat to health.

Nitrogen Dioxide

Thirteen million tons of this gas and related nitrogen oxides were poured into the air of the United States in 1965. Nitrogen dioxide is the main end-product of a series of chemical reactions in the exhaust fumes of internal combustion engines and furnaces.

Severe pulmonary fibrosis and emphysema, that is, scarring and distention of the tissues of the lung, may develop in persons regularly exposed at work to nitrogen dioxide at levels of from 10-40 ppm. What damage, if any, is directly caused by nitrogen dioxide at the typical urban air level of 0.035 ppm is not known.

Nitrogen dioxide is a key factor in the production of the Los Angeles photochemical smog.

Photochemical Smog

This peculiar condition of polluted air is something quite different from the London 'pea-souper', which was a combination of smoke and fog. The basic ingredients of photochemical smog are lots of motor-car fumes, bright sunshine plus windless conditions and a thermal inversion to allow high concentration of the necessary chemicals.

Photochemical smog was first experienced on a large scale in Los Angeles County in California in 1940. This area has all the ingredients, a high population of motor cars, very bright sunshine and regular thermal inversions. A high ozone level (0.15 ppm for an hour) is taken as evidence of serious photochemical smog, and Los Angeles now suffers from it for about 30% of the year.[16] More and more cities are experiencing photochemical smog, though on a lesser scale.

The chemistry of photochemical smog production is extremely complex and not yet fully understood. It is triggered off by nitrogen dioxide absorbing ultra-violet light, which is also responsible for producing tanned skins when sunbathing. This sort of chemical reaction, which occurs because of the effect of light, is called photochemical. Photography is, of course, an example of our everyday use of such reactions. In smog production a whole series of reactions occur in which other pollutants, such as the hydrocarbons resulting from the incomplete combustion of petrol, take part. As a result secondary pollutants are formed, the most important of which are ozone and peroxyacetyl nitrates (abbreviated to PAN).

Photochemical smog causes extreme irritation of the eyes and respiratory illness in some people.

Ozone

Ozone, as we have already seen, is a major constituent of photochemical smog. Ozone is a form of oxygen. Normal molecular oxygen, which forms a fifth of the air, is composed of two atoms of oxygen. The ozone molecules contain three atoms of oxygen; this apparently small difference has profound chemical and biological significance. Ozone has a peculiar characteristic smell, the one you detect after a powerful flash of electricity.

Ozone does, of course, occur naturally at a low level—0.02 ppm—in unpolluted air; in fact, one school of medical thought at the beginning of the century believed that ozone 'revitalised' one and recommended their patients to go to the seaside where it was believed that sea air contained lots of this life-giving substance. One wonders what these gentlemen would have thought of a statement recently made on the toxic effects of ozone: 'The exquisite toxic potency of ozone has been made abundantly apparent not only as an acute, fast-acting pollutant, but as one with long-term chronic insidious potential for all forms of animal life.'[17]

Ozone can bring an increased number of 'attacks' to asthmatic individuals when the level goes over 0.25 ppm. This is not an impossibly high concentration in certain cities; a recent survey (1964-67) in Los Angeles recorded a peak 'oxidant' concentration of 0.65 ppm, and levels exceeding 0.15 ppm occurred on 30.1% of the days surveyed.

Levels of 1.5-2 ppm can produce temporary ill effects within two hours, respiratory effects and reduced mental capacity. Ozone is an occupational hazard to helioarc welders who sometimes get lung damage. Rats and mice exposed to 1 ppm of ozone daily for a year suffer chronic lung injuries. Some scientists now believe that ozone can act as a tumour accelerator, and that prolonged exposure to oxidant air chemicals could accelerate the human ageing process.[18]

Lead

Lead poisoning is one of the oldest industrial diseases known. Lead can severely damage blood cells and has also led to tragic cases of brain damage in children. But until recently many scientists believed that the levels at which lead was found in urban air were too low to worry about. For example, a British Government report, 'The Protection of the Environment', claimed 'the amount (of lead) that is emitted from motor vehicle exhausts is, in this country, trivial. The air in the most congested street contains far less lead for people to breathe than is safely permitted in factories.'[19] The last sentence of this statement contains a dangerous *non sequitur* in that it assumes that permitted levels in factories are in fact safe. However, Russian neurological research shows that such maximum permissible levels are generally far too high. Furthermore, a growing awareness of the manner in which some substances have been found to accumulate along food chains leading ultimately to lethal doses for certain creatures has shaken this complacent view— particularly since there is evidence to show that the airborne load of lead is markedly greater in the Northern Hemisphere than it was in pre-industrial times.

The evidence for this has been obtained by examining the levels of lead in sediments from the ocean bed and arctic snow, which can be dated. On the reasonable assumption that the amount of deposition of lead from the air into the ocean or snow was related to the concentration of lead in the air at that time we can get figures for the levels of lead in the air many years ago.

Even though the global level of lead has increased so rapidly there is much more lead in city air than in rural air. In remote rural areas one finds a few tenths of a microgramme in a cubic metre of air ($\mu g/m^3$); in a typical American city 1-3 $\mu g/m^3$ is usual. This rises at times to 25 $\mu g/m^3$ in Los Angeles, and 44 $\mu g/m^3$ has been recorded in a road tunnel.

The main source of lead in urban air comes from petrol containing leaded anti-knock fluid. Already the increase in atmospheric lead brought about by this innovation in petrol composition is reflected in the level of lead in rain. Other sources are pesticides, and the burning of coal and rubbish. In two American cities it was found that rainwater, under natural conditions proverbially pure, contained a higher level of lead than the maximum permissible level of lead set by the US health authorities for drinking-water (0.05 microgrammes per litre).[20] We shall discuss this problem later in relation to the automobile industry. The removal from the market of petrols containing lead compounds will reduce the amount of lead in the air.

Professor Derek Bryce-Smith has said that nobody knows whether infants and children are more sensitive to lead than adults, but in the light of recent experiences with heavy metals 'it would be prudent to assume greater sensitivity'. The level of lead in human blood has increased sharply over the last thirty years. In Britain studies of blood-lead levels in school-children show the problem to be quite as serious as in the United States. Inhibition of enzymes occurs at blood-lead levels found in city-dwellers and Bryce-Smith claims that: 'No other toxic chemical pollutant has accumulated in man to average levels so close to the threshold for overt clinical poisoning.'[21]

Lead poisoning in children is one particularly nasty result of the pollution resulting from dereliction. The New York City health department estimates that a 'silent epidemic' of lead contamination may be affecting as many as 25,000 slum children, who pick up their daily dose of lead from chipped lead paint in old buildings. Many small children chew non-food objects, and their menu can unfortunately often include lead-painted woodwork. The problem is associated with the poor decaying urban areas. The lead may accumulate over several months without the child showing any untoward symptoms. Treatment with chelating agents—chemicals that bind the lead ion and remove it from the body tissues—has reduced the mortality rate from lead poisoning from 66% to 5%. Unfortunately, 25% of the survivors suffer brain-damage, which can reduce a healthy child to a vegetable.

This sort of poisoning could easily be avoided by removing old lead paint, or replacing the houses. In New York several successful rent strikes have been organised after lead-poisoning incidents to force landlords to take action over this. Professor René Dubos said, '(this problem) is so well-defined, so neatly packaged with both cause and cures known, that if we don't eliminate this social crime, our society deserves all the disasters that have been forecast for it.'[22]

Water Pollution in Urban Areas

The cities use and pollute enormous quantities of water each day. After use such water is often passed into the water courses without proper treatment. The degree of treatment varies from country to country, from strict regulations to a more or less free-for-all situation in some developing countries.

The citizen is directly affected by water pollution in a number of ways. Water, or at least clean water, becomes a dwindling resource and therefore more expensive. Industries which must have clean water either install expensive cleaning equipment or move. Many of our cities are marred because the rivers which run through them are polluted, becoming unsightly and smelly, and robbing the citizens of a convenient recreational amenity.

The most important danger posed to the cities by polluted water is the danger to health. In the nineteenth century the rise of the industrial cities brought very grave public health problems. But in the United States the death rate from typhoid has dropped from 35.8 per 100,000 in 1900 to 2.5 per 100,000 in 1935, and today it is almost unknown in America.

The remarkable decline of cholera and enteric epidemics was secured by purifying drinking water and making sure it did not get contaminated by sewage. The establishment of a public health service for the notification and isolation of infection completed the process of eliminating these diseases. When one thinks of the benefit achieved, the solution proved to be remarkably cheap as well as simple. Whilst that particular battle has been won in the developed countries this is far from the case in many urban regions of the developing countries. Many parts of the world still face public health problems relating to water supply which Western Europe and America solved seventy years ago. The rapid urbanisation has led to an enormous gap between what is needed and the actual potable water supplies and sewage treatment plant. Such cities face a threat from water-borne bacterial diseases such as cholera, typhoid and dysentery as well as mosquito-borne filariasis, schistosomiasis (bilharzia) and virus diseases such as hepatitis.

The public health problems facing the developed countries today over city water supplies are rather more insidious. Very little is known of the possible long-term effects on health of numerous often unidentified compounds that enter sources of water supply. It is accepted that potable water should be free from viruses, from toxic, carcinogenic, and allergenic substances, and from nitrates and nitrites. But the Pasteur Institute of Paris has found all these in the drinking water of large French cities. They found that the River Seine above Paris contained only 15 pathogenic organisms per cubic centimetre whereas below Paris the figure was 1,500,000/c.c. They also found that the River Seine contains 'relatively large' quantities of 3, 4-benzpyrene and noted that this compound is found in most French rivers, especially in estuary mud; 3,4-benzpyrene is, as we have seen, potentially a very dangerous substance.[23]

The urban citizen is not only dependent on the efficiency of municipal water works, but, as we shall see later, on the behaviour of people and industries many miles away and often in other countries.

Solid Waste

How many of us, when we fill our dustbins, ever pause to reflect on the enormous problem raised by solid waste-disposal in cities. Strikes by refuse-collectors have from time to time illustrated graphically just how much rubbish and junk we produce. The two chief arguments in favour of improved waste-disposal are, firstly, that present methods interfere with the quality of life and, secondly, that they are a drain on our resources. In Western Europe, which has a population of about 300 million, 82,125,000 tons of solid waste are produced each year.[24] The USA is even more bountiful in its production of rubbish, producing 190 million tons per year.

If the American people and their solid wastes were spread evenly over the United States there would be in each square mile of the nation 56 people surrounded by 54 tons of rubbish which would include: 3 junked cars, 26 discarded tyres, 8,500 bottles, 17,000 cans, one ton of plastic and $8\frac{1}{2}$ tons of paper.

As the standard of living goes up, so does the production of rubbish. By 1980 it is estimated that each American will produce an average of 8 lb. of solid waste daily against the current 5 lb., giving a grand total of 340 million tons per year. Actually even more waste is produced, for these figures are based only on returns by collection agencies. The true figure may well be 360 million tons per year right now.

Solid waste can be divided into three categories: domestic and trade, industrial, and scrap cars.

Domestic refuse has changed its character. Its volume is growing at an even greater rate than its weight. This is due to the increasing emphasis on packaging materials, particularly plastics; this means that the overall carbon content of solid waste is increasing and there is a temptation to increase the amount of incineration of refuse, which will shift a lot of this carbon from the ground into the air as CO_2. There are some advantages to be had from improved packaging, but a 25% increase in consumption of packaging materials over a ten-year period reflects the commercial pressure to sell goods on their packaging rather than on their quality. A rationally planned society would examine the necessity for all this packaging and seek to control it. A related problem is that of consumer durables. Consumer pleasure is maintained not merely by using the goods but by receiving them. This leads to greater profits for manufacturers —by careful analysis they have learned how to build-in physical obsolescence, that is they have taught customers to accept shoddy products. We are going to be faced with a larger problem than ever before in disposing of furniture, TV sets, radios, refrigerators, washing machines; not merely because there are more than ever before but because people do not or cannot keep them for as long as they used to.

In the United States over 7 million motor vehicles are scrapped every year; in Britain 7 million cars will be scrapped in the next ten years. Not only are the numbers of cars growing, but they have increased power and speed and they wear out more quickly. They have a more complex design, with a wider variety of materials, including ferrous and nonferrous metals, plastics, textiles, and wood, which increases the cost of breaking them down for salvage.

Cars are not designed for ease of disposal. In Europe 10% of old cars are abandoned on the wayside, in open spaces, in woods or on streets. They are a danger to children from broken glass, sometimes a fire risk and always an eyesore. Scrap-yards are filled with hideous mounds of mangled wrecks as a result of their partial break-up for spares.

Car tyres are a major problem in themselves; in America 100 million a year are discarded, of which 70 million are not reclaimed and have to be disposed of.

Most solid waste is treated on disposal sites; in the United States, where there are over twelve thousand such sites, 94% are inadequate, that is they are either uncovered, or cause water pollution, or burn materials with no air pollution control. Some towns burn their waste in giant incinerators—8% of American waste is treated by some three hundred municipal incinerators. They often add to the air pollution problem, for 70% of them have no devices for air pollution control.

A great deal of contract work is done by private solid waste-disposal firms, particularly for industrial wastes. Their drivers are often paid by the ton of material removed, and are sometimes not too careful where they dump it. There was a case in North Wales where a dump regularly caught fire, giving off highly noxious fumes as a result of the chemical interaction of wastes from chemical works, put there by contractors.

In advanced countries current waste-disposal methods do not seem to pose any direct health hazards because of the well-organised public health services, but entomologists have reported that in African towns malaria-carrying mosquitoes breed in water at the bottom of old tin cans. In Calcutta the public services are breaking down and 'all the garbage and all the refuse of living and of workshops ... are dumped, not in stated spots at stated times but everywhere and any time along the streets and lanes'.[25] Such cities become centres of disease.

Noise

Noise is inherent in urban living. It used to be one of its attractions—the bustle and chatter of many human beings living and working together. Unfortunately such sounds are becoming drowned in a cacophonous bedlam of roaring traffic, construction-site racket and screaming jets.

Noise is the most subjective pollutant; sounds that are music to one ear are dissonance to another. The same individual can find a given level of sound bearable or unbearable according to his situation and mood.

A visitor from the past of about twenty-five years ago would be struck not simply by the general increase in noise level but also by its changed composition. The rattling, clanging tram and the chuffing, whistling steam locomotive would have gone, their relatively intermittent noise replaced by the continuous monotonous roar of motor traffic.

Although many cities are in a process of redevelopment, the problem of noise is generally relegated to a minor item in the budget. Thus we find new blocks of flats built virtually without sound-proofing, so that the residents are subjected to the constant barrage of domestic noise. Domestic noise is also changing, consisting not merely of children's games and family rows, but whining, whirling and beating electric appliances and transistor radios.

Noise-level can be measured by other than subjective means. The most commonly used measure is the decibel (dB(A)). A convenient feature of the dB(A) measurement is that the smallest change in sound intensity detectable by the human ear is about 1 decibel. A 'quiet' sound level is 50 decibels.

The dB(A) scale can be confusing, for it shows a whisper as 30, a typing pool as 65, and a heavy truck as 90, whereas we know that typing pools and trucks are more than 2 or 3 times louder than a whisper. This is because the scale is logarithmic, not linear, which means that every 10dB(A) increase gives, roughly, a doubling of loudness. Thus in our example the typing pool would be ten times, and the heavy truck seventy times, louder than a whisper. It is important to remember this feature of the dB(A) scale, in case someone tries to tell you not to worry about the take-off noise from a supersonic jet transport because it is only 5 decibels more than a Jumbo's 115 decibels. This would mean that the S.S.T. is half as loud again as the Jumbo at take-off.

Sound Levels and Human Response

Response Criteria	Sound Source	dB(A)
Threshold of hearing		0
	Broadcasting studio	20
Very quiet	Soft whisper (15 feet)	30
	Bedroom	40
Quiet		50
	Air conditioning unit (20 feet)	60
Intrusive	Typing pool	65
Telephone use difficult	Main road traffic (50 feet)	70

Annoying	Ringing alarm clock	
	(2 feet)	80
	Pneumatic drill (50 feet)	85
	Heavy truck (50 feet)	89
Hearing damage		
(8 hours)		90
Very annoying		95
Maximum vocal effort	Shout (from 6 inches)	100
	Jet airliner	115
	Super-sonic jet	124
Pain		135

(Adopted with modifications from Report of the Council on Environmental Quality, US Government Printing Office, 1970.)

General city noise causes no obvious physical harm, although if it stops a person sleeping this can lead to illness. Prolonged exposure to noise of about the 90dB(A) level—about the loudness of heavy road traffic—can impair hearing. Many people are subjected to this, and worse levels, at work. It is said that up to sixteen million Americans are threatened with hearing damage because of noisy working conditions.[26] We shall have more to say on this matter in the chapter on the factory environment.

Sudden noise can produce changes in the physiological state—speeding up the pulse, rate of respiration, and muscular reactions. This means that the physical mechanisms of the body are thrown into an alarm response, although the person may not be consciously frightened. To be constantly thrown into such a state must be both fatiguing and damaging, leading to hypertension and ulcers in certain individuals. Unfortunately not much is yet known about such effects. There are all sorts of strange phenomena that need explaining. For instance, why is it that musical noise may at times improve people's ability to perform tasks, but at other times, or with other individuals, impair it? How often do accidents at work occur because of noise? How often are they caused because the worker has unconsciously been distracted by noise, or has even been prevented from noticing an auditory warning signal? How much stress, irrational behaviour, and bad nerves are caused by noise? There is a higher incidence of mental illness, as measured by hospital admittances, on the flight paths to Heathrow Airport, London, than in adjacent areas. Is aircraft noise literally driving people mad?[27]

The worst and most common noises are aircraft and road traffic. The indications are that they are going to get worse, and more and more people are going to be disturbed and upset by them. An indication of the scale of annoyance caused by aircraft noise is given by the fact that in 1970, 50 of the 140 major American airports were involved in formal complaints, often law-suits, about noise levels. By 1975, fifteen million Americans will live near enough to airports to suffer from intense noise.

For the next fifteen years, the number of aircraft movements will grow, and the average 'plane will make rather more noise; the geographical area affected by this noise will also increase. After 1985 the rate of growth of aircraft noise may diminish somewhat, as a result of technical advances in 'plane and engine design and the vertical take-off and landing aircraft.

An equally noisy future is predicted for the roads, if the present trends continue. The British Road Research Laboratory has written '... between 19% and 46% of the UK urban population of 45,000,000, live in roads with traffic flows producing noise levels likely to be judged undesirable for residential areas. If the noise characteristics for individual vehicles remain the same as those on the roads at present, and the estimated growth in the number of vehicles occurs, by 1980 the range of population exposed will increase from 30% to 61%. If in addition to the growth in the number of vehicles, the noise emitted by individual vehicles were to increase by 5dB(A), it is possible that up to 93% of the urban population would find the noise undesirable ... these figures all relate to disturbances at home'.[28]

We should not forget, however, that traffic noise can be reduced. The loudness of engine and transmission noise on large trucks could be halved by complete enclosure. This would cost between £50 and £200 per vehicle. Unfortunately, any further reduction would require a redesigned engine. Traffic noise would be reduced if roads were in tunnels or cuttings, but these would cost several times as much as conventional roads. Houses can be sound-proofed if you have the cash. Double-glazing, roof insulation and door sealing costs between £100 and £300. If Britain were to sound-proof all the houses now exposed to too much noise, the bill would come to between £440m and £1,320m.[29]

Ordinary working class people, are, as with all other forms of urban pollution, going to be hardest hit by the growing noise. They live in poor quality houses, in over-crowded conditions, closer to sources of noise—such as factories and main roads. Nor can they afford the high costs of sound-proofing their houses. Furthermore, since they are poorer, they are less able to fight legal struggles against changes such as new roads or airports.

6
The Factory and Place of Work

The working class are exposed to a wider range of damaging pollution, both at home and at work, than any other section of the population. They get more dirty air, more filth and rubbish, more dirty water; they are exposed to more poison and noise than anyone else. Yet none of the official reports of the American or British governments discuss the problem, and their talk of pollution as a non-political problem hides the differential effect that pollution has on the various social classes.

In this chapter we shall examine the manner in which bronchitis strikes the working class and the enormous range of diseases caused by the disgusting conditions still found in some factories and places of work. As I. W. Abel, President of the United Steel Workers of America, has said:

'If coke-making facilities and electric furnaces, for instance, are now judged to be an evident danger to the health of the community, how much more so are they a threat to the health of the workers within the plant ... Conditions in our mills and mines, especially uranium mines, demand immediate relief.'[1]

In recent years the health record of American industry has worsened. Between 1958 and 1969 the injury frequency rate for manufacturing industries (that is the number of lost-time accidents per million employee-hours-worked) rose from 11.4 to 14.8. More than 14,000 deaths and 2 million disabling injuries occur each year in American places of work. On an average day 55 workers are killed and 8,500 disabled, the number of work-days lost because of such accidents is five times greater than days lost to strikes.

Despite all the talk of environmental quality, the Occupational Health and Safety Bill was vigorously opposed in the United States Congress before it was finally enacted in December 1970. The Act is a mild one, the penalty for a *serious violation*—one in which there is a sustantial probability of death or serious physical harm resulting—is a maximum fine of $1,000. If a violation actually results in the death of an employee the offending party only faces a maximum six months jail sentence and/or a fine not exceeding $10,000.

Marx's words, written over a hundred years ago, still apply: 'What could possibly show better the character of the capitalist mode of production than the necessity that exists for forcing upon it, by acts of parliament, the simplest appliances for maintaining cleanliness and health.'[2]

Bronchitis and the British Working Class

In Britain bronchitis is one of the diseases most clearly related to social class. The death-rate from bronchitis increases as one descends the social scale. If the mean death-rate of the British from bronchitis is taken as 100, then the death-rate for people aged 20-64 in social class 1 is 34, in class 3 98, and in class 5 it is 171. (Professional and managerial people belong to class 1, intermediate white collar—class 2, skilled workers—class 3, semi-skilled workers—class 4 and unskilled workers—class 5.) Similarly, hospital admissions for bronchial complaints show a clear class differential. The rates of admission for bronchitis per 1,000 admissions are: class 1—6.4, class 2—6.5, class 3—7.6, and class 5—17.2. There is no escaping the conclusion that 'Bronchitis remains a disease closely associated with poverty, poor housing conditions, and unhealthy environment, particularly in the cities'.[3]

Smoking cigarettes may possibly be an intervening variable. This issue can be clarified by recent studies carried out on children, who do not smoke. We will cite two studies, one on a local scale and one on a national scale.

The first study was made in Sheffield, England.[4] The researchers studied the incidence of bronchial disease amongst school-children selected from three areas with different levels of pollution. The three areas were all working-class areas, with over 50% of the children's fathers being in class 3. One was the Greenhill area, a new council estate with smoke pollution of 100 microgrammes per cubic centimetre per day; the next was the Longley area, an old council estate with a smoke level of 200 microgrammes/cc, and the third was the Attercliffe area of old terraced housing with a level of 300 microgrammes/cc. It was found that chronic coughs were prevalent in all three areas, but worst of all in the smokiest area. The range of chronic cough was from 22% in Greenhill to 50% in Attercliffe. They also found clear class differences. 15% of classes 1 and 2 had chronic coughs, 29% of class 3 and 37% of classes 4 and 5.

The researchers concluded that the area is of more significance in the epidemiology of disease than class. Yet the area in which a family lives is generally determined by its social class; and the working classes tend to live in the worst places.

The national study was of children in the 6-10 age group. The researchers found a clear class differential in the frequency of cases of chronic cough and bronchitis. Among these working-class families, the lower rate was in rural areas and the higher rate in the heavily polluted cities. This could be due to the tendency for working-class children to live in the worst part of the city, whereas the other classes live in the less polluted suburbs. This also holds for adult populations. In rural areas, the male adult death-rate from bronchitis is 110 per million, but in the industrial areas of Newcastle and Bolton, the rate is 204 per million. Colley and Reid conclude that

> 'This large survey of chronic respiratory disease in children has confirmed earlier reports of a social class gradient in this disease among children. Of special import is the fact that this gradient, which is so pronounced a feature of bronchitis mortality in adults, appears in children even before they have begun to smoke. Nor is this social difference disappearing, for it is as clear now as it was in Douglas and Bloomfield's study of fifteen years before... There is thus as little room for complacency about respiratory disease in the young as in those in middle and late life where the national record is unenviable... This study underlines the special urgency of the problem... among working-class families.'[5]

It seems reasonable to assume that a high rate of lung diseases amongst workers is caused by their living nearer to factories than the middle class, thereby suffering the brunt of air pollution. In the Haverton Hill area near Middlesbrough[6] the Local Authority built an estate of 150 houses opposite the I.C.I. works in 1926; the factory soon expanded rapidly, and its proximity, combined with an unfavourable prevailing wind, soon made the area extremely polluted. In 1951 the area was scheduled as category D in the Durham County Development Plan, which meant it would gradually be run down, by failing to replace unfit houses. At the public enquiry, I.C.I. challenged this, claiming that

> '... the houses were of good quality; that the area appeared unsatisfactory only because it contrasted with the open countryside around it; that it was no worse than other industrial areas; that I.C.I. were not the only producers of pollution; and that it had been encouraged to make as much smoke as possible during the war for reasons of national defence.'

By 1959, the situation had become so bad that the Local Authority were considering rehousing the people on the estate;

in view of this a survey was undertaken. The area was solidly working class with 43% of class 3 and 53% of classes 4 and 5. When questioned on their attitudes to the estate, three-quarters of the people said they disliked the dirt, and over half said they wanted to move.

> 'The environment is extremely unpleasant and causes considerable damage to houses, fabrics, and less certainly, health. By stunting plant life and imposing impossible difficulties on housewives trying to keep their houses clean, it encourages a "couldn't care less" attitude. In consequence the area is regarded in Billingham as not only physically filthy, but also as an area in which only "rough" people will live. The residents show their resentment of the physical and social conditions by the fact that over half of them want to leave.'

Diet probably plays an important part in determining how well you survive polluted city air. Apparently vitamins A and E help safeguard the lungs against the ravages of air pollution. It is thought possible that vitamin E protects A from destruction by air pollutants, and vitamin A in turn is crucial for the formation of healthy cells in the lining of the lungs.

In a series of experiments, one set of rats was given a diet which included vitamin E, and another set given the same food with the vitamin E removed. Both sets of rats were exposed to air containing 1 ppm of ozone. The rats receiving vitamin E survived twice as long as the others.[7] The urban poor generally have a far less nutritious diet than those higher up the social scale, and this may be another factor causing the working class to receive more damage than other sections of the populace.

The proximity of working-class houses to factories places their occupants in greater danger than those higher up the social scale. A British study found that on the premises of a steel works, the monthly dust-fall rate varied from 250-500 tons per square mile, and fell off rapidly with distance from the plant to about 50 tons three-quarters of a mile away from the plant. The safe limit is 30 tons.[8] A German study of a steel mill found up to 133 tons per square mile, on a location half a mile from the site.

Most cases of asbestos disease, occurring outside actual 'on the job' exposure, are caused by an exposed worker bringing fibres on his clothing into the home, or through escapes of fibre from the factory into the air of its immediate surroundings. There have been reports of neighbourhood cases of asbestos disease from many parts of the world. A Russian scientist has reported experiences suggesting impaired health in children living downwind of an asbestos plant. In Finland, pleural calcification, a characteristic of asbestos exposure, was found in one out of thirteen adults, farmers and housewives, in a village near an asbestos mine.[9]

Occupational Diseases and Dust Diseases

Many workers contract diseases as a result of pollution at their place of work. Occupational diseases are many and varied and in this chapter we can do no more than indicate some of them.

The effect of dust on the lungs of workers was recognised in A.D. 61 by Pliny, who suggested that workmen protect themselves by covering their faces with transparent bladders. Again, in 1500, Agricola, a physician in the mining town of Joachimsthal, concerned about the 'suppurating diseases of the lungs', advised that air in mines should be purified by ventilating machines and he recommended the use of veils to prevent inhalation of dust. Gradually many more diseases caused by dust were recognised and these were known by the occupation in which they were found, e.g. miner's phthisis, stone-cutter's asthma, grinder's disease,[10] and in 1838 Stratton coined the term 'anthracosis' to denote disease in which the lungs were blackened by carbon. Since then byssinosis (1860), silicosis, asbestosis (1920), and bagassosis have been discovered.

One can classify such diseases according to whether they are caused by a vegetable or a mineral dust; two contemporary problems caused by these are byssinosis and asbestosis.

Byssinosis

The disease byssinosis was first discovered over 100 years ago, but the first registration of its effects was given by the Registrar-General in his report of 1910-12 when the death-rates for the workers in cotton card-rooms showed a significant excess of respiratory disease. One hundred and twenty-six 'strippers' and 'grinders' from Blackburn Mills were examined and it was found that nearly 75% of them suffered from an asthmatic condition due to the dust.

The disease has a characteristic history. At first workers notice pain in the chest when coming into contact with the dust after a short period away from any contact, but the time it takes to recover is usually small. However, it gradually lengthens until eventually there is severe and permanent disability. During the 1920s attempts were made by the manufacturers to reduce the dust. But the Departmental Committee on Compensation for Card-Room Workers (Home Office, 1939) found the disease still prevalent among these workers and in 1940 a compensation scheme was set up for totally incapacitated men who had worked for at least twenty years in dusty rooms. This scheme now includes women, and pensions are payable for 50% disability. The disease is still prevalent.

Byssinosis has now been reported in sixteen countries and up to 40% of the workers in the dustiest parts of mills in the United States and Holland have byssinosis. Other studies have shown similar figures for Northern Ireland flax mills.[11]

Many other, often serious, diseases are caused by vegetable dusts—e.g. bagassosis amongst workers handling sugar-cane.

Asbestosis

In 1907 Murray reported the first cases of pulmonary fibrosis amongst a group of workers who had been exposed to asbestos and subsequently the term 'asbestosis' was coined. In 1930 Merewether and Price showed that asbestosis was a serious health hazard in the asbestos industry and in response to this manufacturers tightened up their safety requirements. In Britain in 1931 the Asbestos Industry Regulations were introduced which prescribed compensation, and limited exposure to asbestos. The asbestos industry has grown very much since then.

The existing regulations have several loopholes. In Great

Britain in 1963, 19,000 people were employed in industries which used asbestos; but the regulations applied to only 12,000 of these, mainly in asbestos cement, asbestos textiles, and brake linings.[12] In many industries such as steel, where asbestos is used in lagging, boiler-making, where it is used for insulation, or motor vehicle repair, where it is found in brake linings, these regulations do *not* apply; and asbestos exposure may also occur in the building industry and in docks. Furthermore, the regulations[13] specifically exclude their application to 'any factory or workshop or part thereof so long as such process or work (with asbestos) is carried on occasionally only and no person employed therein for more than eight hours in any week'.

The number of asbestosis cases is increasing[14] and, according to a recent United States survey, 'regrettably, substantial numbers of persons exposed to asbestos in their occupations have never come under a protective occupational hygiene program'.[15]

Asbestos has also been shown to produce mesothelial tumours. 'It now appears that a worker, or even a person outside a suspected industry, with an apparently low level of exposure to asbestos may die of mesothelioma of the pleura or peritoneum after a latent period of up to forty years and long after he or she has left the industry.'[16]

An even more disturbing finding was Hourihane's report in 1964 that in some cases exposure to asbestos was so slight that it could not be related directly to industrial exposure and only the presence of asbestos bodies in the lungs indicated association of asbestos with the tumour.[17] The significance of this, for building site workers and others, who are exposed to small amounts of asbestos in the course of their job, is not hard to see. Other alarming studies show that South African miners who worked with insulation containing asbestos had a rate of bronchogenic cancer eight times greater than the national average.[18]

Occupational Cancer of the Bladder

The first cases of occupational bladder tumours were recognised in the chemical industry, particularly among workers in the manufacture of synthetic dyestuffs. In 1856 'mauvine'-dye made from aniline was developed. Later colours were synthesised from intermediate derivatives of coal-tar, and of these alpha- and beta-naphthylamine, and benzidine, were later shown to be carcinogenic. Although these were in use on the Continent by the 1890s, they were not introduced into Britain on any large scale until the 1920s. In 1895 Rhen, in Germany, reported four cases of bladder tumour amongst forty-five workers in one dye factory. This is a high incidence for what is a relatively rare disease, and he attributed the cause to aniline (wrongly, as was later shown).

By 1920 over a hundred cases had been reported to the International Labour Office (ILO) with aniline still given as the causal agent, but by then benzidine and naphthylamines were also suspected. By this time the manufacturing process using naphthylamines and benzidine had been introduced into Britain, the United States, Italy and France.[19]

In 1928 there came the first British report of bladder tumours in workers from the chemical industry. In 1926 there was a report from Russia, in 1934 one from America and one from Japan in 1940. During the 1930s pressure grew for occupational urinary cancer to be scheduled under the Workmen's Compensation Act in Britain, but many firms, in the words of Dr R Case, 'thought this step premature'.[20]

Finally in 1948 the Association of British Chemical Manufacturers initiated a survey by Dr Case into urinary cancer. Commenting on the delay, Case says: 'In retrospect it now seems astonishing that the epidemiological evidence that had accrued, although it was fragmentary, should have been so lightly discounted.'

The Case survey was made over the period 1948-53 and it uncovered an extremely high incidence of urinary cancer in workers in various occupations in which the naphthylamines and benzidines, now shown to be cancer-causing agents, were used. In the general male population, the incidence of this tumour was under 1%; but it was 25% in workers exposed to amines, 55% in workers exposed to beta-naphthylamine, and 100% in distillers of beta-naphthylamine! The survey also showed that chemical workers exposed to these chemicals developed tumours considerably earlier than did males who developed such tumours naturally. The workers developed their tumours mostly at the age of 30-40 years as opposed to the normal 55-70 years. On the basis of these findings, urinary tumours were scheduled as an industrial disease in 1953, the manufacture of beta-naphthylamine abandoned and the Scott and Williams 'Code of Working Practice' was introduced. This gives detailed recommendations for the manufacture and use of some compounds and processes which have proved or are suspected to be carcinogenic. The code has been adopted by the International Labour Office. According to Scott 'Its observance should result in the elimination or the reduction to a safe level of contact between operator and carcinogen'.

However, this is not the end of the story. Occupational urinary cancer had occurred in the rubber industry, in which compounds of the naphthylamine type were used. These were used as antioxidants to stop rubber perishing and were withdrawn from use in 1949 as a result of Case's work. However, the use of other antioxidants which were also suspected of being carcinogenic continued and, in the period 1952-65, 65 deaths occurred as against a predicted 30 from the national average. In the words of Case, the Rubber Manufacturers' and Employers' Association decided against a survey in case 'their employees were unduly alarmed'. However, late in 1966 The Industrial Papilloma Committee was set up to co-ordinate a survey of many industries, including the rubber industry. Veys published a recent survey showing that ten workers developed urinary cancer in one factory although they had not been employed in the rubber industry until *after* 1950 when the naphthylamine-derived antioxidants were no longer in use.[21] Case (1966) suggests that experimental evidence shows that rubber curing releases twelve different amines, any one of which may be a carcinogen.

Other industries now also seem to be affected. In the cable industry Davies (1965) demonstrated a high mortality rate in one rubber mill and it is thought that the rubber and mixes may contain a suspected antioxidant.[22] It has even been suggested that the finished product may be carcinogenic and Case, commenting on this, says: 'If this is so the situation could be very serious, for it would mean that we have a hitherto unsuspected class of carcinogenic compounds to deal with, and other methods of making antioxidants might yield a dangerous end-product.'

Another study showed that workers in the retort houses of the gas industry might be running a considerable risk from the beta-naphthylamine which could be produced from the coal tar.[23]

The urinary cancer story is a history of too little and too late coupled with downright obstruction. Dr W. C. Hueper, former head of the Environmental Cancer Section, American National Cancer Institute and a recipient of the United Nations Cancer Award, declared: 'The various parties directly concerned... industrial management, the industrial medical profession, government health and labour agencies, and labour organisations—have revealed too frequently a lack of concern for human suffering and loss of life and a tendency to obstruct measures of prevention and control rarely equalled in other areas of occupational cancer hazard.'[24] Hueper has outlined the kind of legislation required to control the occupational cancer threat; it includes

> The registration of workplaces and other establishments using carcinogenic or potentially carcinogenic compounds. Companies to provide the controlling agency with detailed reports of: the manner in which such compounds are used, the number and types of people exposed to them, the commercial products into which they are incorporated—this will enable potential threats to the general public or workers in other industries to be assessed, the methods of waste disposal—a check against possible air, water and soil pollution.
>
> Strict and regular inspection by trained inspectors.
>
> Regular medical screening of exposed employees.
>
> The keeping of detailed records of employment and health by plant managers, including data on incidents and accidents causing unusual exposure of individuals.
>
> All known or suspected cases of cancer to be reported to the control agency.
>
> Individual employees should have the cancer risks that they face in their work carefully and honestly explained to

them, and should automatically be granted expert legal advice and aid.

Violators of agreed codes of safe conduct should be subject to criminal law—cancer being a deadly weapon.

All data and records of the control agency should be available to scientific investigators *without* the prior approval of the companies concerned.

Agriculture

Life in the open air is not necessarily a healthy one, especially if one is a farm-worker. Farm-workers suffer from a wide range of occupational illnesses. Dust-caused diseases are particularly prevalent, and have been known about for a long time, as the following table shows:

1705—Ramazzini recognised that lung complaints could be caused by dusts from flax and hemp, and since then vegetable dust diseases have been continually recognised.

1860—Greenhow recognised the connection between the dust of cotton and flax in the development of byssiosis.

1913—Collins discovered that mildewed cotton caused weaver's cough.

1928—Baagoe discovered that flour caused miller's asthma.

1932—Campbell discovered that hay caused farmer's lung.

1940—Jamison discovered that sugar-cane caused bagassosis.

1941—Bohner discovered that gumacacia caused printer's asthma.

1956—Lowry discovered that corn and nitrates caused silo filler's disease.

1965—Reed discovered that vegetable dust caused bird breeder's lung.

Farmer's lung develops through the handling of mouldy hay or other mouldy vegetable produce; the causal agent is a fungal antigen in the hay. This illness affects approximately 2% of the general farming population. The attack may last for weeks or months and the subject develops a cough, shortness of breath, anorexia, and anxiety, and considerable loss of weight is not uncommon. Permanent lung damage occurs after repeated attacks.

As we can see from the chronology of agricultural dust diseases, a chronic dust hazard has always existed in agriculture. But over recent years, this danger has become more acute with the introduction on an evergrowing scale of grain-drying plants and silos. In Britain, the National Union of Agricultural and Allied Workers has complained for years to the government about this dust hazard, which is even greater in agriculture than in most other industries. Yet despite years of 'consideration', there are still no regulations requiring the fitting of dust extraction plants in those places. Neither are there any regulations to deal with the hazard of carbon dioxide in silos.

In some parts of the world many farm workers are made ill by the handling of pesticides.

In 1964 in California there were 1,328 cases of sickness directly attributed to pesticides—of these 60% were caused by workers handling them. Agricultural workers who had contact with such chemicals had twice the sickness rate of those workers who did not; the specialist pest-control workers had a rate three times higher. In the period 1951-64 there were twenty-seven occupational deaths from pesticides in California alone. Further, the number of chemical burns received by agricultural workers is increasing. Often these were caused by factors beyond the control of the operator, for instance, 10% of herbicide cases were caused by change in wind direction.

Poisoning from pesticides is very common. In 1964 13 million acres in California were sprayed by 'plane with pesticides. Apart from the risk to people on the ground, among the employees in this industry alone there were twenty-five cases of systemic poisoning from phosphate pesticides. This usually affected unskilled workers who loaded and cleaned the 'planes —about twenty were in this category.

The attitude of the pesticide companies has always been to

play down the dangers. Dr Howard Mitchell, chief of the Bureau of Occupational Health in the Californian Department of Public Health, has been extremely critical of these companies. He points out that his Bureau's studies on the effects of pesticides on health had been hindered, since neither the chemical industry nor the government would release data on the tonnage of the various pesticides used in California.

Dr Mitchell has described a particularly nasty epidemic of organo-phosphorus pesticide poisoning among peach-pickers in 1963. Almost a hundred peach-pickers required medical treatment. The spraying of parathion had been so intensive in certain orchards that the heavy residues on the leaves 'made it virtually impossible for a worker to pick fruit from the trees without being poisoned'.[25]

Tunnel Worker's Disease

An example of extreme physical conditions is found in tunnel-drilling. The classical theory of decompression sickness was propounded in 1878 by Bert. It is due to the evolution of nitrogen gas bubbles in tissues and in the bloodstream, which leads to constriction and damage of blood-vessels. Nitrogen dissolves in the blood under pressure, and on removal of the pressure when the worker surfaces it forms the bubbles.

The conditions of decompression necessary to avoid the formation of bubbles were laid down in Britain in 'Work in Compressed Air—Special Regulations' (Ministry of Labour and National Services, 1958). The table was based on the findings of a study done in 1908. However, in spite of adherence to these regulations, decompression sickness is still prevalent.

Decompression sickness takes three forms:

i) 'Bends', which are cramps in the limbs—these are temporary and not too serious.
ii) Chronic conditions of central nervous, respiratory and cardio-vascular systems. These are fatal unless rapidly treated.
iii) Caisson disease, which is a result of pressure on the articular surface of joints which produces a collapse of

the surface of the joint, lesions, and eventually osteo-arthritis. Caisson disease is permanently crippling and, if contracted, will prevent a worker from being able to work again.

A British study showed that 19% of tunnel workers eventually develop bone lesions as a result of caisson disease. The 1958 regulations are ineffective in controlling this, although it seems that they eliminate the risk of (ii); but the 'bends' are still fairly common and it has been suggested that repeated attacks of the 'bends' may combine to produce the bone and joint lesions. However, this is not yet established for certain.

A survey of Clyde tunnel workers showed that of 241 compressed air workers, 66 showed definite radiological evidence of late or early bone lesions when X-rayed.[26]

Noise at Work

The hazards of industrial noise were first described by Ramazzini in 1713. In his *De Morbis Artificum* he describes how those engaged in the hammering of copper 'have their ears so injured by that perpetual din—that workers of this class become hard of hearing and, if they grow old at this work, completely deaf'.

Modern industrial machinery has greatly accentuated the problem. However, compensation for occupational hearing loss is not usually claimable; if it were made so the cost to industry would be immense because of the scale of the problem. 'The potential cost (of compensation) of noise-induced hearing losses to industry is greater than for any other occupational disease.'[27]

The problem is not simply one of hearing-loss, for noise also produces psychological disturbances in the people exposed to it. In 1940 Vernon showed that excessive noise such as is found in many factories produces 'nervous irritability and strain'. Noise may also produce hyper-activity of the autonomic nervous system. This would tend to produce such results as increased heart-rate and liability to ulcers. The quality of personal life of people exposed to excessive noise is reduced by interfering with communication, at even a conversational level, and by affecting their nervous tempo.

Studies on hearing loss from all over the world have exposed the enormous damage done to the hearing of workers by occupational noise.[28]

In Italy, a survey of the steel industry examined a sample of 743 steelworkers of varying trades; *all* had occupational hearing loss. Another survey examined 103 shipyard workers and found that *every* riveter and caulker among them was affected. A national industrial survey, of 118,668 workers in factories ranging from heavy to light industry, showed that 10% of the whole sample had occupational hearing loss.

In Czechoslovakia it was found that 40% of workers in pump test rooms had occupational hearing loss. A study of heavy truck drivers noted that, in a sample of 51, *all* had occupational hearing loss.

In the USSR of 135 drivers of underground mining vehicles, 91 were affected.

In Australia, a survey of 5,127 workers of all grades showed that 33% had occupational hearing loss.

A recent United States government report estimated that 16 million American workers are threatened with hearing loss.

In France, forge workers have been examined and 45% of the sample had hearing loss.

In Japan, a national survey screened 87,890 workers exposed to noise and showed that at least 10% had occupational hearing damage.

A WHO publication states that 'It is only recently that some machine manufacturers have begun to appreciate the noise problem and to design their products accordingly. Even so, their efforts may be thwarted by technical difficulties, or the prohibitive costs of incorporating sound-reducing devices.' But the manufacturers, like the workers, have always known about noise. It is just that they are only now having to appreciate the growing opposition of those who suffer from it.

Legislation to control noise is hampered by there being no widely accepted criteria for damage risk. There are many types of exposure to noise and indeed many different types of noise sources, and expert opinion on maximum safe intensity levels is therefore anything but unanimous. A Russian survey in 1962 showed that high-frequency noise at 85 decibels produced deterioration of muscular performance and disturbances of cortical and autonomic nervous functions. This suggests that levels should be reduced to 65-70 decibels. 'Many', says the WHO report, 'would consider this limit industrially impracticable.' If the worker is going to continue being damaged, is he going to be fully compensated?

The attitude towards compensation for occupational deafness varies from one country to another. In the United States, 22 states have occupational noise compensation laws, and in California 352 claims were made in 1962, a minute figure considering the numbers at risk. Britain, however, has no compensation for occupational hearing damage. In Norway, occupational hearing damage is classified as an accident, entitling the victim to insurance benefit. In Japan, compensation is established at retirement, if the occupational hearing loss exceeds 60 dB.

Industrialists are very frightened of being faced with legislation which can lead to compensation claims, so they lay great stress on the difficulty of distinguishing between claimable and non-claimable—natural—hearing damage. A WHO report states, 'Industry must be protected against claims it is unable to meet—without some control measures, the potential liabilities of industrialists could be astronomical.'

Technical Innovation and the Working Environment

We have merely sketched a few of the many diseases caused by an unhealthy work environment. Not only are there many more, but since the nature of industry and the compounds and processes used are constantly changing, we may expect to see many new problems arise in the future. Far too few technical innovations are properly examined from the point of view of the effect they will have on the worker. For example, in the foundry industry, self-setting sands have been developed. In order to use these sands, not only do workers have to work faster than they did with traditional techniques, but they are forced to breathe in obnoxious fumes. The self-setting sands give off formaldehyde fumes which cause the eyes and nose to stream, rather as if the worker had a bad cold. Workers are expected to regard such conditions, where the chance of permanent damage seems relatively slight, as part of the necessary conditions of work.

Another innovation, which has caused great suffering to workers involved in manufacturing it, was the enzyme detergent. The detergent industry introduced biological detergents for market reasons. The enzymes used in the detergents caused a severe allergic complaint to develop in many workers who handled it and breathed in the enzyme. The victims complained of breathlessness, and a specialist who examined many of them wrote, 'a striking feature of the histories of those more seriously affected was the statement, by robust and phlegmatic individuals, that they thought they were going to die during the worst phases of the attack'.[29]

Dr M. L. H. Flindt, an occupational health expert at Manchester, believes that 'in addition to causing acute illness, inhalation of this material may lead to irreversible impairment of lung function'. In this case, not only did the detergents industry fail to adequately protect their workers from being injured, but there was not even any technical necessity to manufacture the damaging product.

The legislative process for the recognition of industrial diseases is geared against the worker. The industrialists are quick to innovate, but not so quick to recognise their responsibility for the health of their employees. To get any concrete changes in favour of the worker often requires a growing stack of diseased bodies to be built up over a period of many years.

The tardiness in recognising vibration syndrome as a prescribed industrial disease in Britain is an excellent example of the law's built-in bias in favour of the employers. The use of vibratory tools, i.e. power-driven hand tools such as drills, is increasing, not only in industry, but also in agriculture and forestry. The continued use of vibratory hand tools by a worker may cause him to develop Raynaud's phenomenon, vibration-induced 'white fingers', or 'dead hand'. The vibratory tools damage 'arteries, nerves, muscles, joints, bones and subcutaneous tissue of the upper limbs'.[30] Raynaud's phenomenon is already a compensatable industrial disease in a number of countries. But in Britain, the question of whether or not the syndrome should be recognised as a prescribed industrial disease has been 'under examination' since 1954! An official report on vibration syndrome, published in 1970, carefully explained their difficulties: 'It has been put to us that some workers are obliged, through the incidence of the vibration syndrome, to change their jobs, or to give up overtime working, so suffering a loss of earnings, for which it is argued they should receive compensation.' But, continued the committee, perhaps some workers give up their work with vibratory tools for non-medical reasons, 'because they find the work, which can be very arduous, unsuitable or uncongenial'. The committee seem prepared to risk the workers' health rather than risk employers' profits.[31]

7
The Rural Environment

The continuing growth of the city has required an ever more ruthless exploitation of the countryside. It is simply seen as a supplier of resources, a fruit to be sucked dry with little thought for the next harvest. It is drained of its food, water, power, raw materials and people. The spread of the city and its factories into the surrounding countryside produces a wasteland which is neither country nor city, poor in culture but rich in filth. Its beauty is shattered by mining, quarrying, power-lines, rubbish tips; its water and soil poisoned by factory farming and chemicals, its solitude broken by motorways. All this waste and destruction are not the results of individual insensitivity, but the inevitable results of our social and economic system.

Agriculture and Economic Trends

Agriculture, the main industry of the rural areas, also adds to this destructive process. Land and labour are exploited more and more intensively to give greater yields. There is a trend towards larger and larger farms seeking more and more profitable production, whilst at the same time there is a reduction in the overall area of land devoted to agriculture.

In Britain, over half a million acres of agricultural land have been lost in the last decade. In the Federal Republic of Germany, every year an area of agricultural land the size of the city of Munich is built over with factories, houses and roads. The land-grab to build houses, roads and airports often takes the most fertile land. There has been a considerable reduction in the number of farms in the last ten years. This reduction will continue, squeezing the small man out, for the average farm is too small to guarantee a return on investment equivalent to that of industry.

In the Common Market area there is still a large peasant population; the average farm has an area of only 11 hectares. Sixty-six per cent of the farms have less than 10 hectares of usable land and only 3% have more than 50 hectares. The economic planners of the Common Market want to reduce the number of farms by 8 million, for the present small farms have profit margins that are too small to finance increased investment.[1] They plan to drive these peasants into the already overloaded urban areas. In the Federal Republic of Germany, between 1949 and 1960, 303,900 small farms disappeared, while 40,700 medium to large farms were created. Even in England and Wales, where the peasantry were driven out long ago, the number of farms has dropped at the rate of 6,100 a year in the decade 1957-67, from a total of 367,857 farms to 306,623. At the same time however the number of large farms, that is those over 300 acres, is growing.

Agriculture is therefore experiencing a great decline in the numbers of its labour force. In Western Europe, 42% of the total labour force in 1910 worked in agriculture; by 1955 this was down to 24% and it is still dropping. In the United Kingdom, the number of agricultural workers fell from 730,000 in

1949 to 346,000 in 1967. Within the Common Market countries, it is predicted that the number of agricultural workers will be slashed from its present 10,000,000 (1970) to 5,000,000 in 1980.

In spite of the decrease in the amount of land under cultivation, and the decrease in the agricultural labour force, productivity per man and per acre is increasing. It is clear that the rate of exploitation of both land and labour is being intensified.

Modern farming methods are designed to squeeze the last drop of value from both land and workmen. This can be seen by examining the increasing production figures, and the increased productivity of the workmen.

From 1955-65 European agricultural production rose by 3.3% a year, and the labour force dropped by 4% a year. Thus the *per capita* productivity rose by about 7% a year. The increase in productivity for agricultural workers has been far higher than for industrial workers. In the Federal Republic of Germany, for example, the *per capita* increase in productivity from 1950 to 1962 was 136% for agriculture as against 79% in industry.

The size of the increased yields of cereals in Britain is indicative of the intensity of land exploitation by modern farming methods. Cereal production doubled between 1946 and 1967, even though the acreage sown only increased by 13%. Even greater yields are thought to be technically possible. This kind of increase in land productivity has been made possible by the introduction of higher-yielding grain varieties, increased dependence upon chemical weed control, and the near-avoidance of any crop rotation. This sort of approach puts a tremendous strain upon the soil.

Many farmers are beginning to practise monoculture, that is, they continuously grow the same crop on their land year after year. In Britain this occurs particularly with grain and high-value vegetables. This has important consequences for the soil. Crop rotation was the traditional safeguard for soil fertility, and furthermore it prevented the accumulation of pests and diseases in the soil. Now with the monoculture system, and the break with rotation, farmers are driven to place more and more reliance upon chemical fertilisers and pesticides.

There are certain economic factors which are partly responsible for this break with good husbandry. A major factor is the concentration of ownership of the food processing industry and the growth of supermarket chains. They have put pressure on the farmers to grow their crops in such a way as to guarantee a particular timing, yield and quality, in accordance with their market requirements. In this situation many farmers have found it convenient to grow their crops under contract to certain companies and specialise in particular crops. Thus farmers find that they cannot afford to break their crop programme by rotating with a less valuable crop.

Scientific and technical thought applied to agriculture has decreased neither the sickness- and death-rates of farm-workers nor the disease-rate of crops and soils. The death-rate of farm-workers is in fact on the increase, as is the rate of dust-caused lung diseases. There is little doubt that current practices are leading to an increase in plant diseases. The number of recorded diseases of principal crops has increased threefold since the 1920s. Only a part of this increase is accountable to discoveries of new diseases, previously overlooked.[2] The heavy equipment now put on the land is destroying the delicate structure of the soil, compacting it so that plant roots cannot penetrate. Crops die from lack of nourishment in soils that have been lavished with chemical fertilisers. Last, but not least, as we shall see later, many of these developments increase the pollution problem.

The Pollution of the Soil

The soil may look inert, but healthy soil swarms with life. Any sickness of the soil will inevitably affect the health of man, the higher animals, and all other creatures that form the delicate web of terrestrial life. Soil is the link between past life, present and future life. Bacteria and other micro-organisms break down the remains of past life and pass it on to living life-forms. When these creatures die it is the living world of the soil that ensures tomorrow's basic nutrients. The soil is the heart of life's dynamic resource recycling system. Therefore any assault on the soil is an assault on the whole of life, including mankind.

A wide range of chemical compounds and radioactive materials, which cannot be broken down by bacteria, have entered the soil over the last few decades. So they accumulate and are passed on into the food webs where their effects may range from the immediately lethal to much more insidious long-term effects. An extreme example of soil poisoning occurred at Smarden in Southern England in 1963, when a deadly poison, fluoroacetamide, escaped from a pesticide factory and somehow got into the soil of local fields. Throughout the first half of that year, cows and dogs mysteriously died. Fluoroacetamide is one of the most deadly poisons ever made; a twentieth of an ounce would be enough to kill more than a thousand small dogs. Government scientists called in to investigate these deaths concluded that the only thing that could be done was to scrape up the contaminated soil, and collect up contaminated water from ditches and ponds, and dump it in the sea. Less spectacular events, but equally dangerous, are going on all the time.

Consider the mighty growth of Los Angeles. Each year its speculative builders cover 90,000 acres of rich Californian soil with filling stations, houses, flyovers and motels. But even Californians cannot live on dollars alone; they must get their food from somewhere. Having sold their fertile lands for building, because there is more money in destroying it than farming it, they are now forced to try and farm desert land. To farm this sort of land it must be irrigated, which costs a lot of money. Most of these costs are not provided by the speculators, nor by the farmers but by the Californian taxpayer. To irrigate $1\frac{1}{2}$ million acres of the Central Valley drylands costs $1\frac{1}{2}$ billion dollars. But this is only the first step in a spiral of costs. The next comes when the irrigation itself poisons the land—a result of deposits of salt left by the evaporated water. Of course, irrigation need not lead to this; it is caused by crude, hurried methods which the farmers are forced to adopt in order to stay in business.

Chemical Fertilisers

Fertility is a relative concept; a fertile soil is one in which plants may grow, but obviously some soils can grow more crops than others. All farmers desire a soil which gives the greatest yield for the least effort, and are open to any sales talk that appears to be promising them increased fertility. There is a great danger that the future fertility of the soil will be destroyed for the sake of short-term gains, using techniques directed at a single aspect of fertility. Fertility depends on a whole complex of factors—climate, the original rock from which the inorganic components of the soil derived, what used to be grown on the soil, and what is now grown on the soil, the soil's structure, its water and mineral content, and so on. Modern farming techniques lay stress on mineral content at the expense of soil structure. Yet without a proper soil structure which allows the roots to penetrate easily, provides proper support, is well oxygenated, well drained and cannot be blown away, soil fertility is lost. The depletion of mineral resources by continuous cropping is a swifter process than the degradation of soil structure, so the farmer notices the former more quickly and sees the advantage of chemical fertilisers, whilst tending to neglect soil structure.

The supply of chemical fertilisers is a huge and profitable operation. The American fertiliser industry in 1968 sold about 2 billion dollars worth of chemical fertilisers, weighing around 38 million tons. In Britain, 625,000 tons of inorganic nitrogen were applied as fertilisers in 1968-9, that is, an increase of 150% since 1957-8.[3]

Nitrogen and phosphorus fertilisers may get washed out of the fields into rivers and lakes. Sometimes phosphorus fertilisers have caused problems as a result of chemical reaction with iron, aluminium or calcium compounds already in the soil. Nitrates have been known to percolate directly into the ground water, and contaminate shallow wells. If babies are given water containing more than 8 or 9 parts per million of nitrates they can be made ill or even killed by a condition called methemoglobinaemia, in which nitrogen ions block the transport of oxygen in the blood. At higher levels, more than 50 parts per million, the water becomes unsafe to give to cattle, causing not only methemoglobinaemia but reduced milk production, vitamin A deficiency, thyroid disturbances, abortions and other reproductive difficulties.

If there is too much nitrogen in the soil, sometimes the plants themselves may accumulate very high levels. There have been reports of such plants poisoning people and livestock who ate them. In Germany children have been poisoned by eating spinach. Not long ago in the United States spinach baby-food had to be recalled because it contained too much nitrogen.

It has been found that pregnant women who had eaten large amounts of forced, fertilised and variously processed and frozen foods, were more prone to anaemia. This was particularly noticeable in certain immigrant women who had previously lived on an agrarian diet. Maybe these conditions are linked with nitrate intake.[4] Some agricultural scientists have argued that they can show that continued reliance on chemical fertilisers can lead to decreases instead of increases in production. The Dutch worker P. Bruin summarises this view: 'The improvement and maintenance of soil fertility can only be brought about by the rational use of both inorganic (chemical) and organic (natural) fertilisers.'

Mixed farming agriculture, which maintained the fertility of the soil over hundreds of years, is giving way to factory farming and monoculture. Cows, poultry and pigs, are crammed together inside factory farms and fed on concentrated foods made from grain produced under monocultural conditions.

A formerly integrated system, maintaining soil fertility, has been torn apart into two separate activities. That is why the excreta produced by these animals is now seen as a form of pollution. It has been suggested that farms should have their own sewage works, or be connected to the municipal sewage works. Here we see the absurd logic of current economics; the place for this animal manure is on the land, not in the rivers and sewage-pipes. If it were carefully prepared and spread on the land as manure, we would reduce pollution, reduce the amount of chemical fertilisers needed, and restore soil-structure.

There seems to be a connection between the increasing use of chemical fertilisers and increasing attacks on crops by pests. The East German agronomist, Dr Gustav Rohde, has shown that compost, besides assisting soil fertility, plays an important part in protecting plants from disease. He found that a properly matured compost is unattractive to many soil insects, and that some of the microflora present in the compost feed upon or parasitise the insects. Rohde has also shown that manure which has not been matured, such as fresh night-soil, fresh sewage or urine, may not only contain disease organisms and parasites, but is attractive to plant pests. So it seems that to properly prepare manure will protect not only man but also plants from disease.

Rohde notes that there is a close connection between the amount of chemical fertilisers used and the amount of pesticides used—the more you use the former the more you need the latter, a convenient relationship for the agro-chemical industry. Maybe the relationship is due to the fact that chemical fertilisers, unlike compost, possess no anti-pest properties. Rohde's research has in fact shown that plants receiving an over-dose of nitrogen are more susceptible to attack by insects.[5]

Pesticide Pollution

As agriculture has become more intensive so its pest problems have grown, and a vast industry has arisen to supply and apply these pesticides. There are no such things as pests *per se*, only species that conflict with human convenience, health and profit. Pesticides are poisons designed to kill such species. There are an amazing number of pesticides. In America some 900 active pesticidal chemicals have been formulated into over 60,000 different registered preparations, though only about 3,500 are actually used in agriculture. Annual expenditure on pesticides in the United States has risen from $30-40 million per year in the period 1930-40 to over $850 million in 1969.[6] In 1965 world production was worth about $1,440 million. Production in the United Kingdom in that year was worth $51 million.[7]

There are three types of pesticides: herbicides, insecticides and fungicides. Herbicides produce the highest sales of the three in the States, although a greater weight of insecticides is produced. All three types of pesticides have caused pollution problems. Mercury-compound fungicides have seriously polluted water, and certain herbicides are posing grave problems, particularly through their military use.

The Insecticide Revolution

Before the Second World War the types of insecticides available were inorganic compounds such as lead arsenate, oil derivatives, and natural insecticides such as derris and pyrethrum.

Throughout the 1930s efforts were made to discover synthetic organic insecticides, and with the growing threat of war, which threatened to cut off the supplies of derris and pyrethrum obtained from the tropics, these efforts were intensified. The Swiss firm of Geigy discovered the insecticidal properties of DDT just before the outbreak of war. This was developed and produced on a large scale in the United States and Britain, together with another chlorinated hydrocarbon—hexachlorocyclohexane (benzene hexachloride, BHC). In Germany, also, intensive research was under way, and a big breakthrough was achieved as a by-product of nerve-gas research. Schrader and his co-workers discovered that many organo-phosphorus compounds have insecticidal properties. These two groups of compounds, the chlorinated hydrocarbons and the organo-phosphorus compounds, have proved to be the richest source of insecticides.

During the early years of their use, these compounds were considered ideal. They yielded many great successes, particularly in the public health field; the spectacular elimination of malaria from wide areas is a well-known example. High hopes and enthusiasm were engendered, and Paul Müller received the 1948 Nobel prize for physiology and medicine for his work on DDT.

As a result of this revolution in insecticides, non-insecticidal research in economic entomology was neglected and more funds and scientists were channelled into insecticide development. For some ten to fifteen years most workers in the field adopted the chemical approach to control, and neglected the ecological principle of the ecosystem. A number of workers and organisations saw the dangers of this attitude; for example, in 1947 the Conservation Committee of the American Association of Economic Entomologists stated in its annual report that

'Many entomologists are inclined to adopt a narrow and short-term view of insecticidal control, and do not give consideration to the overall and long-term effects of insecticides. We call attention to the dangers of this restricted view, and urge the expansion of ecological studies by the various Federal, State, and private agencies.'[8]

The Insecticide Controversy

Throughout the 1950s and early 1960s increasing controversy

arose concerning the wisdom of allowing economic entomology to be dominated by the chemical approach. A great number of problems were created by the over-enthusiastic use of insecticides. The major headaches that arose were:

1. Arthropod resistance to insecticides—over 200 pests have developed resistance to one or more chemicals.
2. Secondary outbreaks of pests other than those against which control was originally directed.
3. The rapid resurgence of treated pests, necessitating repeated applications of insecticides.
4. The toxic insecticide residues on food and forage crops.
5. Hazards to insecticide handlers, people, livestock and wildlife due to drifting of sprays or the accumulation of residues in the environment.
6. The increasing cost of control.

Despite the discussion in scientific journals, only a few biologists were determined or fortunate enough to obtain the necessary time and research funds to develop control measures based on an understanding of the ecology of pests. One such pioneer was Dr A. D. Pickett, who in 1943 began studies on the long-term effects of insecticides used in the orchards of Nova Scotia, Canada. On the basis of research on the orchard ecosystem he and his coworkers developed 'integrated' control programmes in which insecticides could be used without crudely destroying the existing natural enemies of the pests.

The populations of most insect species are attacked, and their numbers controlled, by various predators and parasites. If these beneficial species are also killed by the insecticides, control of the pest can only be maintained by repeated and expensive applications of chemicals; sometimes control of a pest may be achieved at the cost of creating a new pest in its place. The red spider mite became a pest in many orchards after the spraying of insecticides intended to control the original pest species. A large number of the most successful insecticides, particularly the organo-chlorine compounds DDT, Aldrin, Dieldrin, are persistent; once they have been applied their residues remain biologically potent for remarkably long periods. In fact, this property was regarded as one of their major assets by many economic entomologists.

Unfortunately, these residues can accumulate in the soil, can be transported vast distances by water, and can be stored in the body-fat of many living organisms. DDT residues have now been found in the most remote regions of the world even in the complete absence of any local source of the compound. In many places persistent organo-chlorine insecticides have caused the deaths of birds, fish, and other wild life.

The possible effects of insecticide residues on human health have constituted an increasing source of concern. Until quite recently it was thought that no section of the population was exposed to more than 10% of the maximum allowable DDT intake. But this overlooked the fact that ingested DDT is excreted with milk. The Swedish National Institute of Public Health calculated that the average breast-fed baby consumed 70% more than the then maximally acceptable amount of DDT. A similar situation existed with the insecticide Dieldrin. As a result the next generation was exposed to the greatest hazard.[9]

There is now a growing interest in the possible sub-lethal effects of such insecticides. Studies have shown that the activities of drug-metabolising enzymes in liver microsomes are markedly increased when animals are treated with certain insecticides. This increase in activity is referred to as enzyme induction.[10] Treatment of animals with inducers (e.g. DDT) of microsomal enzymes affects the behaviour of steroid hormones, which include some of the sex hormones. This may explain some of the observations of declines in the population of some bird species in recent years.

In birds estrogen, a steroid sex hormone, makes calcium available for eggs. It causes more calcium to be taken up from the diet and less to be excreted or deposited in bone marrow. The calcium in the marrow is then transported to the oviduct, where it becomes incorporated into the egg shell. A subnormal estrogen level interrupts this crucial chain of events in the reproductive cycle. Peakall showed that DDT and Dieldrin induce liver enzymes to break down steroid hormones in pigeons under laboratory conditions. If in nature organo-chlorine compounds induce female birds to metabolise their own estrogen we might expect various hormonal disturbances and symptoms of calcium deficiency; abnormal reproductive behaviour such as egg eating, nest abandonment, nervousness, egg thinning, chipping and breaking. All these symptoms have been observed to have increased amongst carnivorous birds since the introduction of organo-chlorine insecticides. This evidence lends strong support to the view that organo-chlorine insecticide residues are a major causative factor in the decline of certain predatory bird species, for example, the peregrine in Britain and the United States, the osprey in the Eastern United States, the Bermuda petrel in the North Atlantic, and the brown pelican in California. At the time of writing some 14 bird species are suspected to have declined as a result of DDT and PCB poisoning.[11] It is not known just how many other species are similarly susceptible; certainly we can expect species and individuals within populations, including people, to vary in their susceptibility to such sub-lethal doses of insecticide.

Public Response to the Insecticide Problem

A number of countries set up scientific committees to examine the situation as the unfavourable evidence accumulated. In Britain an early start was made, and three reports were issued in the early 1950s. But the real public outcry in the United States and Britain did not start until the incidence of massive wildlife deaths around 1960. The debate intensified and became more widespread with the publication of Rachel Carson's *Silent Spring* in 1962. As a result, the government bureaucrats who dominate decision-making in science and technology were forced, in a number of countries, to justify their activities to the nation.

Special investigations into the use of pesticides were undertaken in many countries. In Britain the 'Cook Report'[12] recommended that Aldrin, Dieldrin, and certain other cyclodiene insecticides be banned from all but a few specified uses for which there was at the moment no alternative method available. The scientific evidence on which this decision was based was obtained by the Toxic Chemicals Section of the Nature Conservancy, and also by amateur ornithological societies. It is significant that non-professional organisations such as the British Trust for Ornithology, The Royal Society for the Protection of Birds, and the Game Research Association, played a very important role in uncovering the extent of the damage caused to wildlife in Britain. The British Advisory Committee on Pesticides and other Toxic Chemicals recommended that more restrictions be made on the uses of DDT but was not in favour of a total ban on its use.

In Sweden the National Poisons and Insecticides Board decided in 1969 to forbid:

(a) Aldrin and Dieldrin completely.
(b) DDT and Lindane in household and garden formulations.
(c) DDT for a two-year test period, in all other uses not covered by (b).

The Hungarians banned or severely restricted the use of organo-chlorine compounds after a series of severe fish kills at one of their freshwater lake holiday resorts.

In the United States the problems were on a much greater scale; therefore the struggle between the interested parties was more intense. One of the most important federal investigations was that of the investigating committee chaired by Senator Abraham Ribicoff; the hearings filled eleven volumes.[13]

Their report made some harsh comments on the neglect by scientists of their social responsibilities:

'The committee asked the scientific witnesses for meaningful advice for Congress, but much of the testimony was inhibited by defence of past positions, employee loyalties, and lack of authority. Scientists should do as thorough a job of preparing answers on aspects of research as they do on the technical details of their work. Academic scientists and their administrators should regard the preparation of a popular speech or article with the same significance which they attach to a contribution to their professional journals. The concept of environmental relationships or ecology as a "quiet" subject and "silent scientists" will be of no help in inserting it (ecology) into our own busy life.'

Some scientists faced an extremely hostile reception in their efforts to get more support for ecological studies.[14] Several scientists who had been studying the varying abilities to withstand insecticide contamination of different bird species, ended the paper describing their results with the comment: 'Unfortunately, the project was terminated prematurely due to political problems at the University of Illinois.'[15] They felt that this was due to pressures from people dependent on industrial support for research.

Recently in the United States various Federal and State organisations have stopped or restricted their use of DDT. Furthermore, the struggle between the 'clean environment' parties and the pesticide industry has taken a new turn. The traditional way of dealing with pollution problems in courts has been by means of damage suits, such as that brought by the Lande Government of the Rhineland Palatinate, against the 'unknown persons' who tipped Endosulphan (Thiodan) into the River Rhine in summer 1969. In America the Environmental Defense Fund (EDF) is attempting to transcend this approach with a new legal philosophy. The goal of the EDF is to persuade the courts in the States to recognise, as a matter of equity and constitutional privilege, that American citizens have a right to the cleanest possible environment consistent with the general welfare. Their former legal adviser, Victor Yannacone, used to sum up EDF policy in the slogan, 'SUE THE BASTARDS'. In their campaign they have been able to force certain courts to hear scientific evidence about ecological damage caused by pesticides.

The American pesticide industry is very worried by these activities. The line taken by McLean, chief counsel for the pesticides industry, is that 'Anti-pesticide people are compulsive types who whilst seeking "youth and purity and the simple and primitive" actually feel a loss of physical powers and are preoccupied with the subject of sexual potency'.[16]

If the EDF fails to win its constitutional case in the courts it will reinforce the growing belief of many citizens that industrial demands are decisive. When Arizona banned DDT in 1969 for a trial period of one year, this step was not taken because of possible insidious effects on the endocrine systems of Arizonan citizens, or on their rights to pollutant-free environment, but because of a possible danger to *profits*. The State agricultural industry feared that their products might retain DDT residues exceeding the legally accepted tolerance-levels and consequently be declared *unmarketable*.

Safety First?

Every step the giant chemical companies make is fraught with soul-searching. There is a balance between safety and profit, and to err on the side of the former may be to risk going out of business. If they have any qualms over the safety of a product they will keep it to themselves. On occasions when doubts are aired by outsiders, other scientists or journals, their corporate reflexes oscillate between soft soap and threats.

A most illuminating illustration of this behaviour has been provided by Shell's reaction to criticism of their 'Vapona' Strip. This product was the subject of criticism in articles appearing in *Environment*, an American journal, and *New Scientist* in Britain. At one stage Shell threatened the former with 'legal action',[17] and were said to have threatened to withdraw all advertising from *New Scientist*.[18] In the event, neither of these threats was carried out. The Vapona, or 'No-Pest strip', is a resinous strip that has been impregnated with an insecticide called DDVP (or Dichlorvos). The insecticide is continuously released into the air in small amounts from the strip and will kill any insects flying around the room in which it has been placed. DDVP is an organo-phosphorus compound and exerts its toxic effect by disrupting the functioning of the insect's nervous system. Depending on the dose, it will kill any creature with a nervous system, from fly to man. To kill a human being by ingestion would take 20 Aspirin-size tablets of DDVP. This is not in fact as toxic as many other organo-phosphorus compounds in common use. Consequently many people assume it would be fairly safe to have in the house. However, considering the unusual way in which it is being used, it would seem that feeding tests were not the most relevant way of testing its safety. In fact tests on pigs showed that they at least could break it down in their intestines so that very little was absorbed into the bloodstream.

Toxic material absorbed via the intestines is taken to the liver where it has a good chance of being dealt with by the liver's de-toxifying mechanisms. But this is not the case with a poison that enters the bloodstream via the lungs. This mode of entry is possibly five times as lethal as oral ingestion.[19] So far we have spoken of DDVP as a nerve poison but it may well have other toxic effects. DDVP can cause mutations in certain plant cells and bacteria, and it has alkylating properties which enable it to cause alterations in the structure of DNA. Thus it is possible that DDVP could cause mutations in bronchial and lung cells of people who inhale it. People are more susceptible to mutations at certain ages than at others; the foetus in a pregnant mother, or small children, are more susceptible than adult males. Can we afford to have such people living in rooms whose air contains even minute amounts of a substance that may be a possible agent of cancer, genetic mutations or birth defects?

When the *New Scientist* raised these questions, they demanded that Shell's private research results, which had been presented to official government regulating bodies, should be published so that all interested scientists could examine them.[20] The request was fobbed off with vague promises. *Environment* met with similar lack of success when it tried to get these data published in the USA. Both journals also approached their respective governmental insecticide evaluation committees for copies of the Shell research data, and were informed by the officials that they could only hand it over with Shell's permission. In the case of *Environment,* Shell went to all sorts of lengths to avoid having to hand over the original results, even to the extent of inviting *Environment*'s scientists to a private discussion in the Hilton Hotel, Cincinnati, with scientists brought at Shell's expense from other parts of the United States, Japan, Italy, France, Holland and England, in an attempt to reassure this persistent magazine. It seems, then, that Shell was willing to talk to outsiders about their results, particularly their conclusions that there was nothing to worry about, without actually producing in public the original research results.

A congressional subcommittee which was investigating deficiencies in the administration of federal pesticide regulations showed a certain interest in the manner in which Shell had gained official permission to market No-Pest strip in the USA.[21] It was the US Department of Agriculture that gave its approval to the product, in 1963. However, at this time the Public Health Service was not particularly happy about the use of such compounds in homes. In 1965 a Public Health Service Committee expressed the view, 'it is unsound to expose man routinely to pesticide concentrations as large as those which kill insects.' The Public Health Service agreed in June 1967 to withdraw its objection to Vapona's continued registration, provided that its label bore the statement 'Do not use in nurseries or rooms where infants, ill or aged persons are confined'. After abortive attempts to change the wording to 'do not use in rooms *continuously* (my emphasis) occupied by infants and infirm individuals', Shell finally agreed to the Public Health Services cautionary wording. However, a survey made by *Environment* in November 1969 showed that there were still large numbers of No-Pest strips on sale that did not carry this warning.

Not every company that mislabelled its pesticide vapour strip was given so much time to sort itself out. The Pesticide Registration Department forced the Aeroseal Company to withdraw its strip four days after receiving a complaint about its mis-labelling. The complaint had been sent to them by Shell. Since Shell were themselves fighting a stubborn rearguard action about labelling, why should they report a fellow company? It transpires from the minutes of the Congress subcommittee that Aeroseal used to buy Vapona from Shell for marketing in their own packages. Later they stopped buying Shell's product and began manufacturing their own strips. The subcommittee found the contrast between the several years' delay allowed to Shell and the four days to Aeroseal most peculiar.

This was not the only battle Shell fought against official demands that cautionary statements should be added to their labels. They struggled for several years in a vain attempt to avoid having to label No-Pest strip with a statement that it shouldn't be used in places where it might contaminate food.

As a result of their experiences, *Environment*'s investigation team concluded that they had been given 'a sense of what it must be like to be a regulatory agency dealing not just with one multi-million dollar chemical company but with many such corporations ... it is a reasonable question to ask whether any regulation at all is possible in such circumstances ...' Today, Vapona strip labels do contain cautionary statements, but the question of whether they should be allowed to remain on the market is still debatable.

Alternatives to Insecticides

If insecticides have created such difficult problems, should we not ban their use? At the moment the answer to this question is, no; insecticides still provide a major weapon in the armoury against pests. Certain insecticides, particularly organochlorine compounds, are being phased out. However their replacements, organo-phosphorus insecticides, may create new ecological problems for they are generally broad-spectrum, highly toxic and short-lived. This, together with their generally non-scientific use, leads to the destruction of natural control agents and repeated sprayings. This leads in turn to resurgences, new pests and a rapid selection for resistance. Therefore the solution to the 'pesticide treadmill'[22] is not simply to replace a few compounds. A more scientifically based approach must be generally adopted.

The work of Pickett in Canada is part of an increasing effort to escape from the emphasis on single cures and to evolve an 'integrated' approach to pest control. This has been described as: —

> 'A pest population management system utilising *all* suitable techniques either to reduce pest populations and maintain them at levels below those causing economic injury, or so to manipulate the populations that they are prevented from causing such injury. Integrated control achieves this ideal by harmonising techniques in an organised way, by making techniques compatible and blending them into a multi-faceted, flexible system.'[23]

The more important bio-environmental control techniques are: —

Biological control:	using predators and parasites of pests.
Microbial control:	using virus, bacterial, and fungal diseases of insects.
Cultural control:	crop rotation, choice of sowing time, ground clearing, drainage, ground breaking etc.
Resistant crops:	development of genetic or nutritional techniques of increasing crop resistance to pests.

Autocidal techniques: control by radiation or chemically induced sterilisation.
Interference with pest behaviour patterns: using pheromones and chemical attractants or repellents.
Physical methods: using radio frequencies, infra-red radiation, visible and ultra-violet light and ionising radiation.

The disenchantment with insecticides as a 'cure-all' has led to increasing research into bio-environment controls.

One of the more recent concepts is the sterile male technique. Dr E. F. Knipling thought of this method in 1937, when he noticed that female screw-worm flies would mate only once. He realised that if sterilised males could be obtained they could be used to reduce the population of the screw-worm pest. The method involved breeding vast numbers of male flies and sterilising them by gamma radiation or by chemical sterilants. These sterile males are later released, and compete with normal males for the attention of the females. In this way many sterile unions are formed; for example, if the ratio of sterile to fertile males is 1:1, then the reproductive capacity of the pest population is reduced by 50 per cent (assuming that the sterile males are as sexually competitive as the normal males). By a series of such releases, involving millions of sterile males, it is possible to get a pest to eradicate itself, hence the term 'autocidal control'. Using this technique the screw-worm was eradicated in Florida in 1958. The method is now being tried in a number of countries besides the United States, and its originators have great hopes that it may contribute to the solution of some major pest problems.

Not all insects can be cultured, and hence there has been an intensive search for chemical sterilants which can be used to sterilise both males and females of pest populations in the field. In this case, instead of relying on mass releases of cultured sterile males, a combination of chemo-sterilant and insecticide might be sprayed. There are obvious dangers in such an approach. Alternatively, chemo-sterilants could be placed in traps to which the pest would be lured by attractant chemicals.[24]

Much research has been done in an attempt to find ways of reducing the quantity of insecticides used in control programmes. One interesting line of investigation has been the use of a combination of attractant and insecticide. An experiment conducted on the Pacific island of Rota used bait consisting of small fibre-board squares treated with an attractant, methyl eugenol, and an insecticide, Naled. These were scattered by an aeroplane all over the island. Fifteen 'drops' over two months resulted in the eradication of the Oriental Fruit Fly from the island by the use of 3.5 grams of insecticide per acre.[25]

The concept of attracting insects to their deaths has been given a great boost by recent work on pheromones. These are chemicals produced by insects which affect insect behaviour. It is claimed that a knowledge of the pheromones of the major insect pests could enable us to interfere with their behaviour in such a way as to prevent them finding mates, or the host plants on which they feed. However, the study of pheromones is still in its infancy.[26]

The Costs and Benefits of Research and Development

The production of pesticides (including fungicides and herbicides as well as insecticides) is a major industry. Pesticides have proved to be very profitable commodities, and the giant chemical companies have been prepared to invest large sums of money in research and development. Galley estimated that the commercial R. & D. budget was £35,000,000 annually. In 1962 American commercial pesticide laboratories employed 650 graduate scientists, of whom 530 were employed by chemical firms. Annual US sales at that time were around $400,000,000.

A similar ratio of sales/R. & D. expenditure seems to hold in Britain, where the industrial R. & D. budget totals £2-3 million for a market worth £22 million.

The cost of developing a new insecticide is very high, and is increasing, partly in response to the more stringent biological testing now required before an insecticide can be registered. Over the period 1956-63, R. & D. costs for a typical insecticide increased from $1,196,000 to $2,918,000, and to $4,096,000 by 1969. During the same period the average success rates of compounds screened dropped from one successful insecticide in 1,800 chemicals screened to one in 3,600. No doubt in response to this change over the last five years or so, the number of patents and registrations granted for new insecticides has dropped.

The Union Carbide Chemical Company claim that up to 1963 they had spent $2,505,000 in R. & D. on their insecticide, Sevin. They had also invested $3,250,000 in plant, giving a total investment of $5,755,000 in a single compound. Obviously the development of new pesticides can be carried out only by the largest chemical corporations.

The old type of persistent, wide-spectrum compound such as DDT is no longer acceptable, and the chemical companies are faced with having to develop highly selective compounds. However, despite advances in the study of the physiological effects of insecticides, the 'tailor-made' insecticide designed for a specific pest still remains very much a pipe-dream. Some industrialists have said that research costs for such compounds would be prohibitive, particularly since, by definition, they would have a more limited sales potential than the conventional insecticides; and if governments insist on having them they should pay for the necessary R. & D. by putting out contracts as they do for military research.

Unfortunately, no figures are available for the amount of money spent on bio-environmental control R. & D. on a world-wide scale. If the United States' experience is typical, then a decreasing percentage of government funds is going to pesticide R. & D. The US Department of Agriculture's (USDA) Entomological Research Division spent more than 60 per cent of its research budget on pesticide-related research during the 1950s. Today the figure is between 20-25 per cent, and the percentage going on bio-environmental control research has increased correspondingly.

However, despite the improving situation, bio-environmental control R. & D. is still much worse off than pesticide R. & D. In 1964 the research budget of the USDA Entomological Research Division was $10,576,000 (this included $2,891,000 on conventional insecticides). This is only about a quarter of the amount spent by industry on pesticide R. & D. (State institutions probably spend as much as Federal bodies, but even allowing for this, the balance in terms of research funds and facilities available is heavily in favour of chemical methods.)

Industrial research contributes very little to bio-environmental control. In 1963-64 only twenty-one man-years were devoted to the field by private laboratories in the States, out of an estimated total national research effort in bio-environmental control of four hundred and forty-six man years.

Since it is quite likely that success in bio-environmental control, as in other fields, is closely related to the energies and funds devoted to research and development, it is worth looking at the benefits already derived from a selection of successful ventures.

Californian figures indicate that successful bio-environmental control is cheaper, as well as safer, than pesticidal control. The original Californian biological control triumph, the introduction of the Vedalia beetle in 1888 to combat the cottony cushion scale *Icerya purchasi*, saved the citrus fruit industry at a cost of only $5,000. In the period 1923-1959 biological control in California saved $115,000,000. If partially successful schemes are included this figure can be increased to at least $125,000,000.[27] These savings were achieved at a cost during the period of $4,300,000 on research and application costs, giving a return of $30 for every dollar invested. It has been estimated that in the USA pesticides give a return of $4-$5 for every dollar spent. Clearly, biological control has proved to be about six times as cheap as pesticidal control.

A similar success story is reported by Dr Simmonds of the Commonwealth Institute of Biological Control (CIBC). He claims that they have achieved seven successful introductions of parasites which led to permanent control. The cost of these controls was only £25,000 and yielded an average return of 100% per year. Over the period 1928-66 CIBC spent a mere £1½ million and their current annual costs are around £200,000 a year. Thus their expenditure over the whole of their existence has been about the same as the cost of developing a couple of modern insecticides![28]

Other bio-environmental control techniques may be more costly than biological control but they still show a very high return on the original research investment. It was mentioned earlier that in the South-eastern States the screw-worm has been controlled by the sterile male technique. This pest attacks cattle, and used to cost the livestock industry in those parts some $20 million annually. The cost of the research programme for developing the sterile approach to control was

$250,000 over the period 1950-57. The cost of the eradication programme was $10,000,000 over eighteen months, that is, about one half of the expected annual losses due to the screw-worm. It is estimated that over the period 1960-64 the livestock industry benefited by $100 million (less eradication costs). For every dollar invested in control there was a saving of $10, and on the original research costs there was a return of $400 for each dollar spent.

8

Fresh Water Pollution

Water availability is a shaper of civilisation and cultures. Two river valleys of the Middle East—the Tigris and the Euphrates—were the cradles of civilisation. The ready availability of water is essential for society. By poisoning the rivers and lakes of the world man poisons civilisation.

Without water there would be no life as we know it. Sixty per cent of a man's weight is water. Some organisms can contain up to 95% water. The formation of large bodies of liquid water containing dissolved chemicals such as methane and ammonia is thought to have been the key event which made life on earth possible.

The fundamental biological role of water depends chiefly on two properties. Firstly, more substances are soluble in water than in any other liquid. This makes water an ideal medium for chemical reactions, which occur more readily in liquids than in gases or solids. Secondly, hydrogen and oxygen play a fundamental role in biological systems, and water, a molecule of which is made of two atoms of hydrogen and one atom of oxygen, is able to provide these elements to living systems in an easily assimilated form.

Since there is an enormous amount of water on the earth, why worry about polluting it or running short of it? Unfortunately most of the planet's water cannot be directly utilised by us. For a start 98% of all water is salt water, and most of the fresh water is locked away in the Polar ice caps. It is estimated that man has available to him 350,000 cubic kilometres (km^3) of surface fresh water, 150,000 km^3 of ground water (water underground), and 13,000 km^3 of atmospheric water (which can be ignored for practical purposes until it reaches the surface of the earth). However, in spite of the fact that this is an immense amount of water, water is becoming a scarce commodity, because of increased demands and because much of it is being rendered useless or less useful because of pollution.

In the United States the water shortage situation is even further advanced. There, the per-capita consumption of water was 1,200 m^3 in 1963. By the year 2000, it is estimated that American per-capita consumption of water will be 1,500 m^3 per year.[1] The increase in demand is due to changing domestic and industrial needs. For example, as more families own bathrooms and washing machines so domestic consumption will increase. Industry uses enormous amounts of water; it takes 1000 litres of water to make 1 kg of silk, 900 litres to treat five litres of petroleum, 200 litres to make 1 kg of paper. Of course, improvements in processes may reduce the amounts of water required. In old steel works 300-400 m^3 of water were needed for each ton of steel made, but in a modern plant the figure is 10 m^3. However, in spite of such improvements industry's water needs will grow enormously in the future.

The availability of water supplies can have an important effect on economic development. In 1966 a Council of Europe report stated:

> 'In many densely populated and highly industrialised areas of Europe, the water situation can already be described, without exaggeration, as critical. Not only are there supply

difficulties, but the quality of the water is deteriorating. In the Eastern and Northern *départements* of France, in the Ruhr basin of the Federal Republic of Germany, in Belgium and in the Netherlands, the problem is assuming disturbing proportions and is already threatening economic development.

Are there any large untapped sources of fresh water? Obviously we cannot yet increase the total amount of water on the planet, but amongst the more grandiose schemes for obtaining more fresh water are tapping the water locked away in glaciers and polar ice caps, artificially induced rain and desalination of sea water. For most parts of the world, however, such schemes are well into the future. The current shortages are not such as to justify the costs or risks involved, although the two latter suggestions have been tried on a restricted scale.

United Nations experts claim that:

'For the time being, the only way to increase water resources is through a better use of the natural cycle of water in the liquid state, coupled with: efforts to reduce the loss of water through evaporation; sound river management to ensure conservation of flows which would otherwise be wasted; more energetic pollution control measures.'[2]

In countries without proper control measures, the extent of industrial pollution can be absolutely staggering. For example, in a FAO report[3] we hear of a small '*torrente*', a river with a small summer flow in Italy, that receives effluents from 1,109 industrial establishments over a length of 47 kilometres

The world is moving into a very critical situation in which more and more water is being polluted at a time when more and more water is required. The tremendous economic consequences of these problems of increased water requirements and increased pollution have caused a growing number of countries to promulgate legislation, and to create the necessary administrative machinery to enforce it. In Britain where such changes are probably most advanced there is evidence that water pollution, after growing for most of this century, has been checked and is getting less severe. However, there is a danger that having stemmed the polluted flow the government may become complacent.

The argument against complacency has been clearly put by the World Health Organisation.[4] Assume a low rate of increase in water consumption, say 4%, which leads to a doubling of demand every twenty years. If the proportion of water requiring purification remains static then water treatment capacity has to double every twenty years. Since the nearest and cheapest water sources are tapped first then the costs of additional supplies will be greater. The volume of polluted water will increase, requiring proportionate increases in treatment plant. Even if we assume that the river volume remains the same the increased amount of effluent will diminish the diluting effect of the river. If standards are not to fall then the degree of treatment given to effluents must increase, and therefore costs will soar. In actual fact the increased demands are more likely to reduce the river volume, making things even more difficult. Say the WHO, 'when all these factors are taken together, it is clear that even a well-developed country, with adequate authorities, good intentions, and no lack of money, must view the future with a certain misgiving.' As we shall see, such a well-endowed country is the exception and the overall world situation is extremely bleak.

What is Polluted Water?

We must be quite clear about what is meant by fresh water pollution and understand how such pollution is recognised. Dr Key, a British expert, has said:

> 'A river may be considered to be polluted when the water in it is altered in composition or condition, directly or indirectly as a result of the activities of man, so that it is less suitable for all or any of the purposes for which it would be suitable in its natural state.'[5]

Here is another definition: 'Pollution is a *natural* or *induced* change in the quality of water which renders it unusable or dangerous as regards food, human and animal health, industry, agriculture, fishing or leisure pursuits.'[6]

Pollution is classified as mechanical, chemical, organic or mixed. Depending on intensity and frequency it is said to be 'massive' (i.e. a spectacular form with lots of dead fish and so on), or 'chronic'. Chronic pollution may be associated with a long-term reduction of oxygen concentration, the presence of toxic substances, or scarcity of nutrients, which reduce the stream's fitness for life. Under such conditions the fish migrate and less attractive fauna and flora appear. Chronic pollution can be transformed into massive pollution by such factors as rising temperatures or reduction in river flow, which cause the amount of oxygen to drop to an even more lethal level without there being any increase in effluent discharge into the river.

Mechanical pollution is due to solid wastes from such sources as mines, quarries, tanneries (hair and other residue) and paper pulp mills. One can see mechanical pollution, for it is solid suspended matter. The extent to which a river or stream can tolerate it depends on its speed of flow; a slow-flowing stream will be more seriously affected. The chief effect of this sort of pollution is to cut down the amount of light penetrating the stream, thus reducing plant growth and, in turn, the oxygen which is produced by plant photosynthesis. Small solid particles may also irritate the sensitive skin on the inner edges of the gills of fish and cause enough excess mucus secretion to choke the fish.

Chemical pollution generally results from the discharge of industrial effluent. Industrial effluent contains industrial products or by-products which are of no economic value, or whose concentration is too low to allow them to be easily removed. The range of chemical pollutants of rivers and streams is enormous. The most common are: mineral and organic acids and their salts; alkalis; chlorine; metals, particularly zinc, copper and iron; ammonia and ammonium salts; phenols; cresols and cyanides. Such pollutants create toxic conditions which harm the health of creatures that live in the water, or kill them.

Organic pollution is caused by effluents containing nitrogenous compounds and hydrocarbons. Such compounds can be broken down for food by microscopic plants and bacteria. This process requires oxygen which is taken from the water by the microorganisms. This is turn lowers the oxygen level in the water, making it increasingly difficult for fish to breathe. An increase in this putrification process beyond a certain point will cause the fish to migrate, if they are able, or even kill them. In severe cases of organic pollution all the available oxygen can be used up and the putrefication process is then continued by microorganisms which do not require oxygen. These include anaerobic bacteria which produce hydrogen sulphide as the end product of the process. This gas is responsible for the 'bad eggs' smell associated with stagnant rivers. Some organic pollutants act as fertilisers, causing the water to become clogged up by plant life—this variant is called eutrophication.

The chief sources of organic pollution are: domestic sewage (which may also contain dangerous bacteria and viruses); agricultural processing plants (dairies, for example); farms, particularly those with intensive farming units; sugar mills, distilleries, breweries, soap works, tanneries, paper mills, textile factories and dye works.

To sum up, water is subjected to a range of pollutants which in the main exert their deleterious effects by poisoning the water or by reducing the concentration of oxygen, or both.

There is no single measure of the degree of pollution. Many indices are used, some biological, others chemical. Often a mere glance at or smell of the river is enough to give some indication of its degree of pollution. But for really accurate measurements a well-equipped laboratory is necessary. The numbers and composition of the types of plants and animals living in the water form the basis of certain assessment methods; the presence or otherwise of fish and their behaviour may prove particularly valuable indicators. The types of bacteria that are present is extremely important if the water is to be drunk. Generally, however, a series of chemical tests are used which include: measures of the amount of dissolved oxygen (D.O.); the biological oxygen demand (B.O.D.), which is a measure of the level of oxidisable organic pollutants; pH, the degree of acidity or alkalinity and salinity, and so on. Physical factors such as temperature are also noted.

The Extent of Fresh Water Pollution

The water pollution situation in different countries is almost infinitely variable because of the wide divergences in geographical situation, population and degree of industrial development. In this section we shall present an overview of the situation in a selection of countries representative of the whole.

The Council of Europe produced in 1966 a special report which summarised the situation in *Western Europe*. They estimated that in France 1/30 (6,000 Mm^3) of all the fresh water discharged to the sea is now polluted; and that some of the Central European lakes (Morat, Baldegg, Halwill, Zug, among others) have undergone 'disastrous' biochemical changes which have spoilt not only their fishing, but also the usefulness of their water for domestic and industrial purposes. In Lake Constance 'the effects of pollution have reached an alarming stage'. The water of the lake used to be so pure that it could be drunk without first having been chlorinated. Today its bacterial level is so high that drinking water has to be chlorinated. The types of fish found in the lake have changed for the worse. Lake Geneva is one of the largest and most beautiful lakes in Western Europe; but the Report claims that pollution in the lake is now severe enough to pose a threat to fishing, potable water supplies and bathers.

The giant rivers of Europe, often the sources of potable water to the cities on their banks, are more polluted than ever before. Of course, the fact that many of these rivers flow through different countries creates extremely difficult control problems. We shall examine these later on.

The River Rhine, reports the Council of Europe, 'has become, over a stretch of 250 km, from Lake Constance to the Netherlands, a gigantic open sewer'. Bacterial counts taken at various points in the Rhine's journey from its source to the sea show clearly how it becomes increasingly infected by bacteria. At the Swiss canton of Grisons the bacterial count is 30-100 bacteria/cc, at the entrance to Lake Constance 2,000 bacteria/cc, at Kembus 24,000 bacteria/cc and eventually the count rises to 100,000-200,000 bacteria/cc.

It is not only bacteria that accumulates as one travels down the River Rhine. Mineral salts accumulate, particularly chlorides, some 30,000-35,000 tons of which reach the Netherlands section each day. This creates enormous difficulties for the population of Germany and the Netherlands in the lower reaches of the Rhine, for whom the Rhine is an indispensable source of drinking and irrigation water. It is worth recalling that the

Rhine had severe local pollution problems even before the industrial revolution. The poet Samuel Taylor Coleridge (1772-1834) wrote:

> 'In Köln, a town of monks and bones,
> And pavements fanged with murderous stones,
> And rags, and hags, and hideous wenches,
> I counted two and seventy stenches,
> All well-defined and separate stinks!
> Ye nymphs that reign o'er sewers and sinks,
> The River Rhine, it is well known,
> Doth wash your city of Cologne,
> But tell me, nymphs, what power divine
> Shall henceforth wash the River Rhine?'

However, despite the sewage, the Rhine used to have salmon and even at the end of the nineteenth century 100,000 a year were taken from the lower reaches of the Rhine in the Netherlands. There are said to be a few survivors today, but they are apparently inedible because of their foul taste.

The changing level of salmon catches is an excellent indication of the rate of pollution. The fishermen on the River Tees in North-East England took 8,000 salmon annually at the beginning of the century; in 1920 the figure was 3,000, in 1930 2,000, in 1937 23, today none! The last salmon to be caught in the River Thames was in June 1833. In France it is said that in 1900 the River Seine between Rouen and the estuary contained 50 species of fish, whereas today there is 'nothing...except a few diseased eels'. In Switzerland there are said to be an average of 200 fish 'kills' per year destroying 'millions' of francs' worth of fish. Of course one does not always need to kill the fish to kill the fishing. On the River Elbe fish are still 'fairly abundant' but off-flavours caused by pollution render them inedible. As a result, the two hundred professional fishermen working the Elbe in 1946 had dropped to 12 by 1963.

Few countries have truly accurate pictures of the extent of river pollution and so it is not possible to give an average European figure for the percentage of polluted rivers. However, these can sometimes be obtained for the odd country. For example, a FAO report states that 15% of the total length of Austrian rivers are 'heavily polluted' and a further 15% are 'marginally polluted'. In Poland 35% of river length and 6% of lakes are polluted to the extent that fisheries are affected. In France 7% of river length is polluted. In Hungary 7% of river length is polluted and a further 12% is said to be of 'doubtful quality'.[7] A British survey in 1970 found that 952 miles (4.3%) of rivers were 'grossly polluted'; 1,071 (4.8%) were of 'poor quality requiring urgent improvement'; 3,290 miles (14.7%) were of 'doubtful quality needing improvement' and 17,000 miles (76.2%) were 'unpolluted or recovering from pollution'.

There are some bright spots in all the gloom. Some of these are in Britain where the anti-pollution legislation is now beginning to show results. The River Thames had 41 different species of fish in its lower reaches in 1968. Not so long ago the lower reaches of the Thames were often totally deoxygenated—this is now a much rarer event. Control techniques and their success are something we shall come back to. For the moment, as we continue 'gloom-mongering', it is worth bearing in mind that technical solutions to these problems are available.

Not only capitalist Western Europe suffers from pollution. A United Nations report in 1963 listed the following as the chief areas of pollution in the USSR—the reaches and tributaries of the Rivers Volga, Oka, Kama, Don, Dnieper, Dniester, Northern Dvina, Northern Donets, Southern Bug, Kura, Kuban and the Irtysh. The worst polluted rivers are small rivers in industrial regions, e.g. the Miass, Chusovaya, Iset, Tagil and Neyva in the Urals industrial zone and the Krivoy Gorets, Krynka, and Kaimius in the Donbas industrial zone. We shall devote a special section to the peculiarities of pollution in the USSR later.

All regions of the United States have water pollution problems. The most seriously damaged areas are in the North-Eastern States and the Great Lakes.[8] The problems in the North-East are a result of large scale urbanisation and industrialisation, in an era of virtually no pollution control.

Very often regional water pollution problems are created by their particular industrial or agricultural specialisation. In the Mid-West and South-West, farmyard animal wastes and 'run-off' from irrigation schemes and fields carrying salt, fertilisers and pesticides are major problems. For example, water flowing back into the River Colorado from irrigation schemes carries enough leached-out salt to cause the river to become increasingly saline along its course. Pesticides have caused 'fish kills' in Kansas and Missouri. In Appalachia and the Ohio Basin the chief problem is acid drainage from abandoned mines.

It was the discovery of the extent of pollution of large areas of the Great Lakes which shocked Americans into an awareness of how much damage had been done to their supplies of fresh water. The Great Lakes are the world's largest store of liquid fresh water, having an area of 6,000 square miles in a watershed of about 3,000,000 square miles.

A whole series of factors, both natural and man-made, are changing the Great Lakes. The forests in large tracts of the watershed are being chopped down and converted to agricultural land. This leads to increased soil erosion which in turn increases the sedimentation rate in the Lakes.

Vast urban and industrial centres have been built in the region. Other factors are: erosion of beaches, a natural geological process, effects of natural biological changes, natural changes in temperature and rainfall; and increased use of chemicals for agriculture, industry and home use.

The Great Lakes region has been the centre of a hundredfold population increase in 150 years. In the early nineteenth century the whole of the Great Lakes Basin supported less than 300,000 people. By 1966 30,000,000 people lived on the shore or near to the Great Lakes, i.e. about one in three Canadians and one in eight Americans. In the Lake Erie Basin there were in 1966 10,400,000 people on the United States shores and 1,400,000 on the Canadian side. In the Lake Ontario Basin the respective populations in the USA and Canada were 2,300,000 and 3,800,000.

The major urbanisation zones are Cleveland, Akron, Lorain (Lake Erie), Windsor, Detroit, Flint (Lake Erie), Toronto, Hamilton, Buffalo (Lake Ontario), and Chicago, Milwaukee (Lake Michigan).

The increasing use of the Great Lakes as waterways puts yet another great strain on their biological purity. Their use as waterways dates back to the early explorers and fur traders. Since then, much money and equipment has been invested to make the Great Lakes one of the most sophisticated water transport systems in the world. In 1959 the St Lawrence Seaway was opened and the steady increase in traffic has created a pollution threat because of discharges by ships, wastes and potential cargo spills, particularly oil, which can result from accidents or collisions. The degree of pollution varies from lake to lake.

The water quality of Lake Superior is said to be excellent; dissolved oxygen is at saturation point and the transparency 'is phenomenal'. This lake has been protected partly by its vast size but primarily because it only supports a very low population along its sides. The waters of Lake Huron are of good quality still, but some eutrophication occurs in the Saginaw Bay area.

In Lake Michigan pollution is generally restricted to shore areas situated near harbours and urban areas. It is primarily caused by nutrient enrichment causing excessive algal growth. The problem of eutrophication is one of the chief water quality concerns in Lake Michigan.

Howard Johnson, a Michigan State Official, believes DDT residues were 'the most probable cause' of mortalities of coho salmon fry, DDT up to 19 ppm having been found in fish.

Lake Erie and Lake Ontario are the smallest of the Great Lakes but they support the largest concentration of population and industry and they are the most polluted. Their condition is very serious indeed. Professor Barry Commoner has said that 'instead of Lake Erie forming a waterway for sending wastes to the sea, it has become a trap which is gradually accumulating in its bottom waste material dumped into it over the years—a kind of huge underwater cesspool'.[9]

There are disagreements about just how far Erie is from total disaster. 'Lake Erie has its pollution problems but its epitaph is premature.'[10] All the experts agree that Lake Erie is polluted along its shorelines and that there is severe algal bloom in the lake's shallow western basin.

Because Erie and Ontario are the smallest lakes but are surrounded by half the population in the Great Lakes region, they are subjected to the greatest use and pollution pressures by industry and by domestic sewage.[11] This area has abundant mineral and ore supplies, low cost water transport, petroleum, natural gas, timber and hydro-electric power, and this has led to its development as a major industrial area. It is also a large recreational area for fishing, boating, and swimming.

Lake Erie shows significant increases in the amount of dissolved calcium, sodium, potassium, chlorides and sulphates. Such dissolved inorganic substances rose from 140 ppm in 1900 to nearly 190 ppm in 1960. These increases have occurred since 1910 when the major acceleration in population started. Chloride and sulphate are major ingredients of industrial and human sewage. There is bacteria contamination in local areas, mostly on the United States shores of Lake Erie and the Canadian shores of Lake Ontario. Nutrients, particularly phosphate, have created an advanced state of eutrophication, particularly in the western basin of Lake Erie. Accelerated eutrophication is also occurring in Lake Ontario and its effects are being carried into the St Lawrence River. Ontario shows an increase in phosphate level comparable to Erie's. There are two major sources of phosphate; 60% is from household sewage and 40% is due to fertiliser run-off. Other sources of pollution are the dredgings from lake harbours dumped in the lake, refuse, oil spills, radioactive materials, and warm water from power plants. Oil and gas well drilling, which has now started in the lakes, is considered a potential source of pollution.

A major tributary of Lake Erie is the Detroit River which receives very large quantities of pollutants from the heavily industrialised Detroit metropolitan area, and partially-treated sewage from the city of Windsor, Ontario. The Maumee River at the western end of the lake and the Cuyahoga River at Cleveland discharge significant amounts of pollution into Lake Erie.

The Federal Water Pollution Control Administration Report on Lake Erie states that municipal waste is the main cause of pollution. It ranks the local cities in the order of amounts of waste discharged. Detroit, Cleveland and Toledo are the first three villains but Detroit produces more waste than all the other cities combined.

Industrial pollution is particularly great in tributary rivers and harbours. Some 360 industrial sources are responsible for 87% of the total waste-flow into Lake Erie and its tributaries, about 9.6 billion gallons a day of effluent. Electric power production accounts for 72% and steel production for 19% of this flow.

The Ford Motor Company at Dearborn and Monroe, Michigan, is the chief producer of industrial waste, says the report. It discharges 19.7% of the industrial waste (excluding electric power station producers). Next in the rankings are Republic Steel (Lorain and Cleveland, Ohio, and Buffalo, N.Y.) with 14.9% and Bethlehem Steel (Lackawanna, N.Y.) with 13% of the total volume of industrial effluent discharged, excluding electric power producers.[12]

Eutrophication

It is frequently said that Lake Erie is 'dead' or 'dying'. On the contrary, the problem is not death but an overabundance of certain forms of life. There is too much algal growth. As we said in an earlier chapter, there is in all lakes a degree of natural eutrophication but at a very slow geological time-rate which is barely perceptible in a human lifetime. However, pollution speeds this up. Records going back to 1850 show that eutrophication has occurred largely since 1910, when the main population growth started.

Eutrophication is caused by fertilising nutrients from municipal sources—particularly detergents—industrial wastes and land drainages. Phosphorus and nitrogen are major causes, but trace elements play a part. Phosphorus is thought to be the controlling material, much of it washed into rivers from agricultural land. Eighty-nine thousand tons of total phosphorus were contained in fertilisers applied to land in the Lake Erie basin in 1966, and if only 2% of this got into the lake it would represent a substantial input.

Eutrophication and its excessive algal growths interfere with water supplies, aquatic life, aesthetics, industrial uses, recreation, and shoreline properties. Changes in fish populations have been reported; valuable species have decreased and been replaced by less valuable ones.

The enrichment of Lake Erie is not uniform. It is very intensified in the shallow western end but not significant in the deep eastern end. The reason for this is that the western end is much warmer and receives more than half the nutrient input to Lake Erie. The area of the western part of Lake Erie classified as 'polluted' increased from 26% to 100% between 1930 and

1961, whilst the area classified as 'heavily polluted' increased from about 2.5% to 26%.

The elimination of phosphorus from detergents would be a great help in reducing the water phosphorus level. It is thought that 70% of phosphorus in sewage in the USA and 50% in Canada arises from phosphate-based detergents.

Changes in the distribution and abundance of bottom fauna indicate great changes in the oxygen level of the lake bottom over the period 1930-1961 in the western area of the lake. The tubificid worm, which thrives best when oxygen concentration is low is now nine times more common, constituting over 80% of the bottom fauna; whilst the burrowing mayfly, Hexagenia, once so abundant that it used to create a nuisance by being squashed in vast numbers on the roads, is now almost extinct since it cannot live in low oxygen conditions.

Eutrophication causes problems for the fishing industry. Lakes with clear water (oligotrophic lakes) are the source of the most valuable fish—trout, char, chub or lake herring and white fish. Eutrophication causes their progressive replacement by less valuable forms such as bass, perch, and pike and eventually by even less-valued carp and sun fish. Are these changes in fish species due to worsening water pollution or poor fish management? Many biologists such as Kormondy[13] believe the explanation is directly tied to shifted environmental factors that have resulted in conditions in which desirable species fail to reproduce. Although total fish production in the lake continues to be about 50 million lbs, it is of the less-favoured table forms—sheepshead, carp, yellow perch and smelt.

Lake Erie, unlike Lake Michigan, does not yet have a DDT problem. The fish from Erie contain 1 ppm DDT compared to Michigan fish with 13-19 ppm DDT (the FDA limit on fish for sale is 5 ppm). However, what its fish lack in DDT they now make up for in mercury. Fish in the St Clair River-Lake Erie water system now contain such high levels of mercury that state and Canadian governments have either banned fishing or warned against eating fish from these waters.[14] This particular problem will be discussed in depth later on.

So Lake Erie is only 'half dead' and its pollution saga is not fully completed. Barry Commoner believes that a massive problem will soon arise there, for 'most of the inorganic products released into the lake as a result of waste treatment do not flow out of Lake Erie into the sea, but are reconverted into organic matter; much of it remains in the lake, where it forms the huge demand for oxygen that has been so disastrous for the lake's biology ... (and) great mounds of algae have been washed up on beaches'. Algae dies off and accumulates on the lake bottom.

The bottom of the lake has now amassed vast amounts of organic matter. This organic mud is covered by a skin, the main component of which is ferric iron (iron in its oxidised form). Without this skin the oxygen-demanding organic mud would be brought into full contact with the water and remove its oxygen. But for this protective skin to survive, the water must contain enough oxygen to keep the iron in oxidised form. Without oxygen the ferric iron is converted to its ferrous (reduced) form. Ferrous iron is soluble and the once insoluble protective skin could dissolve away. There is, say some scientists, evidence that this has begun to happen during recent summer periods when the lake's oxygen level fell drastically. If the bottom were to 'break', the lake would suffer a massive, acute and disastrous reduction in its oxygen level.

It is feared that some of Erie's biological changes might spread disease. In western Erie the coliform bacteria (Escherichia coli) count rose threefold from 1913 to 1946. But no change has been noted in the eastern section of the lake. Escherichia coli is used as the index organism for aquatic bacterial pollution. Many Erie beaches have been closed because the coliform count is higher than the minimum safe level. In 1967 out of 83 lakeside beaches in Michigan, Ohio, Pennsylvania and New York, 27 were unsafe for swimming all season, 28 for part of the season. The way to solve this problem is to increase the number of sewage treatment plants, and this is going to prove expensive.

The Niagara River is the greatest contributor of pollutants to Lake Ontario. The Niagara drains already polluted water from Lake Erie and receives more waste materials from heavily industrialised Buffalo and Niagara Falls City. Rochester and Toronto contribute major amounts of nutrients to Lake Ontario.

The cleaning up of the Great Lakes will not prove to be an easy task. If pollution stopped completely it would take twenty years at natural flow for 90% of the wastes to be cleared from Lake Erie and Lake Ontario, and hundreds of years from Lakes Superior and Michigan. That is the legacy of seventy years of the private pursuit of profit and public neglect in the Great Lakes Basin.

Water Pollution in Developing Countries

The vast majority of the world's nations are classified as 'developing', that is, they do not have a predominantly industrialised economy based on commodity production. Most of Asia, Africa and South America are underdeveloped in this sense. There are even parts of Southern Europe and North America (the Arctic regions) which are underdeveloped.

However, the label 'developing country' hides a wealth of diversity which is often relevant to the question—how bad is pollution in the developing countries.[15] They may be tropical with high rainfall, or desert with little rain. They can be countries of short, fast-flowing rivers or massive, slow-flowing rivers like the Nile. They may be maritime or landlocked. They might have a very small population density like Zambia or a high one like India. The degree to which they are urbanised differs widely, as does their industrial development. Therefore any answer to our question must perforce be rather sketchy and produce generalisations needing to be treated with reserve. We shall examine examples of the types of problems being faced in a selection of countries.

The extent of pollution is related to the degree of industrialisation, that is, the stage of 'development' reached by the particular country under consideration. Three stages on this path to 'development' may be distinguished.

(a) Predominantly pre-capitalist economies with the bulk of the population engaged in peasant agriculture and little or no industry except for mining. In such countries pollution is of the traditional type, i.e. human sewage. This carries with it the hazard of epidemic diseases due to faecal contamination of water supplies. Zambia is an example of such a situation.

(b) Countries with some localised industrialisation, but with a majority of the population still largely unaffected by the new industries, for example India. Such countries are in a particularly difficult situation for they are faced with increasingly varied forms of pollution by industry before they have adequate resources to cure even the 'traditional' pollution. Furthermore, the growth of industry is leading to increasing urbanisation which will intensify the traditional form of pollution caused by sewage disposal.

(c) Countries with a relatively advanced industrial base, e.g. Rumania or Argentina. At this stage of development the problems become so aggravated that abatement measures become very necessary if economic growth is not to be damaged by pollution. Such countries are approaching the sort of pollution situation reached by developed countries some time ago but they probably lack the wealth to carry out control effectively.

We will make a brief survey of the world and look at the extent of water pollution problems.

In Brazil officials admit having water pollution problems. The main problems are found around their largest cities—Sao Paulo and Rio de Janeiro.

India is a particularly interesting case. At a WHO Seminar Indian speakers spoke of 'serious and widespread pollution... increasing at an alarming rate'. Rivers that were until two decades ago comparatively free from pollution are said to have 'phenomenally changed' for the worse, and 'most of the rivers near urban localities are today grossly polluted'.

Even parts of the big rivers are being polluted, where previously dilution would take care of most waste-matter entering them. Though, of course, the worst problems are to be found in small rivers in industrial areas which are being polluted by a variety of industrial wastes, as well as domestic sewage. R. S. Mehta described the chaotic situation in which 'Almost all factories discharge their wastes without adequate treatment excepting some public sector industries which have installed treatment plants. These industrial wastes have created a serious problem... there is absence of proper legislation both at the federal level and at the state level to prevent this pollution and local bodies have been completely ineffective to curb this.'

The water pollution problem is complicated by religion in India. The Hindus hold certain rivers sacred and periodically gather at fairs and festivals so that they may bathe in the sacred rivers such as the Ganges. The largest festival fair, the Kumbh held at Allahabad, is reputed to be visited by 6 million people on peak days. The Hindu ritual is not completed by merely bathing in the rivers; the faithful must drink a small amount of the water through cupped hands. At Varanansi Benares in Uttar Pradesh state, where numerous holy fairs are held, 71 drains enter the Ganges in a five-mile stretch. No wonder Indian water pollution experts worry about the future.

For over four thousand years the River Nile has enabled millions of people to survive in what would otherwise be a giant sand-patch—Egypt. Egypt is the Nile. Under the revolutionary government it is witnessing an unprecedented growth in population and industry. The new textile, dyeing and finishing, food, chemical and mineral works all produce effluents which 'are generally discharged into the water-courses including the Nile'. There are strict regulations but as an official diplomatically put it to a recent WHO seminar: 'However, the circumstances attending economic development and industrialisation have led to the disposal of most wastes without adequate purification, thus exceeding the permissible limits. Mostorod drain has actually ceased to function owing to the discharge of various untreated industrial wastes. Its water turned black and had such an offensive odour that the inhabitants of the neighbouring villages were disgusted.'

At the moment such problems are localised and the Nile has an enormous flow, but the building of the Aswan Dam has affected the ecology of the delta and the Eastern Mediterranean, a thousand miles downstream. It is fair to say that unless future industrial developments are strictly controlled even the Nile could be dangerously polluted.

Bulgaria and Rumania are two of the most underdeveloped countries of Europe. However, under the rule of the Communist Party they have both experienced an extremely rapid rate of development. Rumania's current economic growth-rate is claimed to be second only to Japan's. What has happened there is almost a classic case of pollution development concentrated into a period of less than two decades. From little or no pollution both countries moved rapidly to a situation where they experienced a good deal of pollution which has made some of their rivers offensive, toxic to fish and unsuitable for irrigation.

In Bulgaria at Dimitrovgrad noxious discharges from a chemical plant killed 175 tons of fish in 1966. Industrial effluent from Pernik has put a thousand hectare irrigation system out of action. In 1967, of some 800 industrial establishments in Bulgaria only 36 had purification plants of their own and only 220 had access to a purification plant. As recently as 1964 there was no legislation. The Bulgarian government plans to have all 'major' sources of pollution under control by 1980.

Japan cannot be called an underdeveloped nation—after all, it is now number three in the world industrial rankings. But in terms of pollution control it is extremely backward and underdeveloped. In 1965 only 77 cities had sewage treatment plants, serving a mere 11% of the population. No doubt today the figure is higher but unless there has been an historically unprecedented building programme the Japanese level of sewage treatment is still surprisingly low for an industrialised nation.

Most of the country is faced with the problem of disposing of 'night soil' from privies. But the uneven development of capitalist progress can produce weird effects. Formerly the simple and obvious way to dispose of 'night soil' was to spread it on the land. Today, however, the chemical industry provides more and more fertilisers for the Japanese farmers, fewer of whom are prepared to take 'night soil'. Of course sewage and 'night soil' treatment plants did not keep pace with the problem and 'night soil' disposal has apparently become one of Japan's most difficult environmental sanitation problems. Consequently there are numerous small outbreaks of water-borne diseases such as typhoid fever and dysentery.

Japanese landscape gardeners always made brilliant use of the local waterways, but the water pollution now threatens all this. For example, the Sumida river, Tokyo's most famous river, used to be clean and beautiful and people were able to enjoy themselves along its banks. Today it stinks of hydrogen sulphide and nobody plays there any more. Industrialists and municipalities do not treat their pollutants unless made to by legislation;

Japan's are no different from Europe's or America's. Faced by weak legislation or poor enforcement they will do little to alleviate the damage they cause to other people.

In Japan the standards established by the Industrial Wastes Law are effective only for effluents discharged into 'designated water areas' prescribed by the Water Pollution Control Law. The 'designated water areas', although they comprise the most important water areas of Japan, do not include many small streams, lakes, bays and harbours. These remain uncontrolled and there is now considerable pollution in such areas.

Once again the fishing industry has suffered in consequence. Saburo Kato told the New Delhi WHO seminar:

'From ancient times the Japanese people have been fish eaters, and rivers, lakes and the sea have constituted good fishing grounds... Water pollution has changed this; some fishing grounds now have a reduced yield, some have been destroyed and some are dangerous. It is no longer rare to find that fish caught in the areas near chemical plants are uneatable because of bad smell or toxicity. Some fishermen have therefore given up their occupation, and others go far away to the sea near Africa to catch fish.'

Even our quick survey of the extent of fresh water pollution leaves little doubt that not only is the situation very bad, but that it is going to get far worse for some time to come. Countries like Britain that are beginning for the first time in fifty years to reverse an adverse trend will prove to be the exceptions.

The Effects of Fresh Water Pollution

Water pollution can damage an enormous range of human activities. Water is used in so many facets of our life; for drinking, washing and cooking, as a source of food, for irrigation, and in many industrial processes.

The quality of water required by industry varies widely according to the type of process. If the water is needed for cooling purposes then it doesn't need to be of a very high quality providing it contains nothing which will corrode or create deposits on the heat exchangers. On the other hand high-pressure boilers require very pure demineralised and de-oxygenated water.

The carelessness or indifference of one firm may cause great inconvenience to another. That is why it is an oversimplification to claim that industrialists are necessarily always against pollution control. A recent court case demonstrates this point. A waste disposal firm polluted a tributary of the Macclesfield Canal in Cheshire. As a result I.C.I. Pharmaceuticals, who drew out water lower down the canal for cooling and sterilising purposes, lost a batch of drugs worth £1,000.

We have already seen the extent to which pollution has affected fresh-water fishing all over the world. We must remember that the toxicity of any pollutant to fresh water fish depends on a wide range of factors; the species of fish, their previous history—are they healthy, have they had a chance to be acclimatised?—the concentration of the pollutant, environmental conditions, temperature, dissolved oxygen, pH, and hardness of water. Coarse fish are generally able to tolerate lower oxygen concentrations than are trout; that is why pollution hits the most highly prized fish hardest.

Riversides and lakesides have always attracted man, not only because of their utility, but also because of their beauty. Some of the most expensive house sites in the London area are on the banks of the Thames in areas like Richmond. No doubt many of the early efforts to obtain water pollution control originated as an aesthetic response to the repulsive stink of the untreated sewage which poured into the rivers running through large cities. 'I counted two-and-seventy stenches,' wrote Coleridge of Cologne. Not only the nose is offended. Some of the sights of polluted rivers can upset sensitive souls—floating faeces, used contraceptives, sanitary towels, suspended sediment, crazy-coloured dyes, old bikes and other rubbish and algae triply offend sight, smell and touch.

People love to bathe in rivers; that is why there are strict classifications of the suitability of water for bathing, based on coliform bacteria counts. People want to use and enjoy their rivers and lakes and pollution should not be allowed to prevent them. But we must also remember that recreational activities themselves can be in turn a major source of pollution.

In order to cope with the growing population and the loss of fertile land more and more land is being brought into production by means of irrigation, and more and more land already farmed is being irrigated to intensify production. Already in many places polluted water is being used for irrigation, thus spoiling the crops or threatening the health of people who eat the crops. The pollution of a river whose waters are removed downstream for irrigation could therefore prove most expensive to human health and crops.

Many of the world's navigable waterways are extremely polluted. Difficulties have arisen as a result of tarnishing and corroding metal surfaces and the blackening of paintwork by hydrogen sulphide. However, only rarely is navigation interfered with, for example when a canal catches fire. The most recent and tragic example of a canal burning was on the Manchester Ship Canal when a mile-long stretch of the canal burst into flames, killing five people crossing in a ferryboat.

The general effects of infection by water-borne disease were dealt with in the urban environment section. The Minamata case will be developed as a special study in the next chapter.

9
The Minamata Disease and Methyl Mercury Poisoning

The most infamous case of mercury pollution in recent times occurred in Japan, where over a hundred people were poisoned. The event showed not only the tragic results that can be created by certain toxic pollutants but also the reckless human activities which intensify such situations.

At Minamata in Japan the Chisso Corp Chemical factory made vinyl chloride and acetaldehyde. The effluent from the plant contained organic mercury compounds. The effluent was run into Minamata Bay over an eight-year period from 1950 to 1958. The fish and shellfish in the bay accumulated the mercury, over 40 ppm of mercury on a net weight basis being found in some shellfish.

Many of the local people ate fish and shellfish caught in the bay and some soon began to show symptoms of poisoning. In the first 11 months of 1956 over 42 poisoning cases were reported. The owners of the factory disputed the claim that they were in any way to blame and the only action taken was to ban fishing in the bay. Nevertheless, it appears that fish and shellfish from the bay continued to be eaten and over the period 1953-1960 a total of 111 people were poisoned.

Some of these people died from resulting secondary infections. The major toxic action of methyl mercury is on the brain. The resulting brain damage shows itself initially in the following symptoms, which normally take 1-2 months to appear. First fingers, toes, lips and tongue go numb. Later a drunken unco-ordinated walking gait develops. Vision is blurred. It becomes increasingly hard to speak or hear clearly—in fact, these latter symptoms make it so difficult to communicate effectively that some of the victims were thought to be insane and were put in mental hospitals! In severe poisoning the speech is loud, explosive and unmoderated. The patient finds difficulty in separating voices in conversations involving several persons. Headache, fatigue and irritability are initial symptoms of which the cause may go undetected. Trembling hands, slurred speech, numbness of limbs, kidney damage and deafness result from severe poisoning.

Authorities of the American National Communicable Disease Centre in Atlanta state that 10% of any mercury ingestion goes to the brain and that 'we might lose only a few brain-cells with each tiny amount of mercury, and the impact might not be noticeable until we are older. Possibly, some persons considered senile are suffering from long-term mercury poisoning.'[1] Industrial mercury poisoning was recognised in the felt-hat industry in the nineteenth century; hence the term 'as mad as a hatter'.

One of the most tragic aspects of the affair was the congenital damage to babies, of which 19 cases were found. The poison was able to cross the placental membranes from the mother to the foetus. In many cases the mother herself showed no symptoms of poisoning; it seems that the foetus concentrates the mercury more than the mother does. Methyl mercury has been shown to cause genetic changes in root cells from the plant Allium and in the fruit fly *Drosophila melanogaster*. However, little is known about the possibility of such genetic effects in man.

The chemical works at Minamata stopped pouring its effluent into the bay in 1958; instead it was put into the Minamata

River. There it proceeded to poison fish and shellfish. It was not until 1960 that a treatment plant was built and even this only gave a partial treatment which did not stop all the organic mercury from entering the stream. At last, after almost sixteen years of public poisoning, the mercury content of Minamata fish and shellfish dropped when in 1966 efficient anti-pollution control equipment was fitted. The Minamata affair is surely one of the most tragic and criminal cases of water pollution.

The fertiliser plant at Minamata provided most of the industrial jobs in the area and workers feared for their jobs. Not until 1968, in response to outside pressures from university and government scientists, did the factory-owners officially acknowledge that they were responsible for the Minamata disease. The national pressure on the factory-owners was increased by a second outbreak of the poisoning disease at Niigata where 120 cases of methyl mercury poisoning led to five deaths and 26 people seriously affected.

Mercury Pollution in Sweden

The discovery[2] that the environment of Sweden was widely contaminated by mercury was a major factor in making that country more environment-conscious. Birds, farm products and fish were all found to contain unusually high concentrations of mercury in the early 1960s. The drastic decrease in the numbers of certain wild species of birds was shown to be due to the use of methyl mercury fungicide seed dressing. Furthermore, many aquatic creatures were contaminated by mercury in the form of methyl mercury, which was found in all types of fresh water and marine fish examined during the mid-1960s. Any mercury level above 0.2 ppm in wet tissue is 'high' and probably unnatural. Levels of this order and higher were reported from many areas, the highest level being 9.8 ppm in a pike caught in the Stockholm Archipelago.

The mercury in the water can be accumulated and concentrated by organisms along a food-chain, and the all-over concentration factor from water to pike is of the order of 300. Mercury is found 'naturally' in fish but is mostly below 0.1 ppm, and always well below 0.15 ppm.

Nobody knows for certain how fish take up mercury. Soluble mercury compounds could be taken up through the gills, and through the food-chain from primary bottom feeders to the predators.

Robert Ehrlich, a geologist at Michigan State University, believes that industrial pollution experts have underestimated the pollution menace of metals such as mercury because they ignored the process of 'organic complexing'.[3] It had previously been believed that the mercury would react with inorganic ions in the water and sink to the bottom in a harmless form. For example, an official of Tenneco (Houston, Texas) argued that his firm's mercury would not cause much trouble in the Houston Ship Canal because the water contained 'a high concentration of sulphites and sulphides. Mercury would precipitate out as sulphide or sulphite salts and stay in the channel-bottom silt because these salts are extremely insoluble'.[4] But if organic molecules are present in the water, and they can be provided by sewage and other decaying animal and plant matter, the mercury can react with them to form soluble organic complexes. As a result the metal spreads through the rivers and lakes more rapidly than might be predicted from laboratory experiments.

Dr Ehrlich and his colleagues have found that much calcium and most copper forms such complexes in rivers and lakes. 'There has been no previous accurate determination of water pollution by elements forced into solution by such organic complexing... common analytical methods do not test for the complexing action,' says Ehrlich. He adds that in lakes and rivers organic complexing can be the dominant chemical reaction for 'the great variety of dissolved organic matter presents an enormous number of potential interactions with metals and other inorganic elements in the water... although organic complexes break down after four days or less, a steady influx of new complexes can be maintained because of the constantly increasing amounts of sewage and nutrients being dumped into lakes and streams.'

A report in *Chemical and Engineering News* adds, 'Dr Ehrlich warns that if enough of the complexes are formed, vital chemical reactions can be stopped, with serious consequences for plant and animal life in rivers and lakes.'

Industrial effluents usually contain mercury in the form of

inorganic divalent mercury, elemental mercury, phenyl mercury, methyl mercury, or an alkoxy-alkyl mercury. Once in a river it can be converted to the particularly toxic methyl mercury. It has been shown that micro-organisms can convert inorganic mercury to methyl mercury, therefore any pollution by mercury compounds which can yield inorganic mercury must be suspected as a possible source of methyl mercury. Göran Löfroth has summarised the major sources of mercury pollution as:

1. Pulp and paper mills which use phenyl mercury acetate (PMA) which may escape in waste water. Since 1966 Sweden has restricted its use.
2. Chlorine factories which use mercury electrodes cause both air and water pollution. It is estimated that eight factories in Sweden lose 25-35 metric tons of mercury a year. The firms are attempting to cut these losses but, says Löfroth, 'they seem reluctant to switch to the equally economic and technologically feasible diaphragm method which does not use mercury'.[5] The mercury cell is essentially a tank containing elemental mercury over which saturated brine, pumped up from underground deposits, is passed. An electric current is passed through the brine from a carbon anode to the mercury which acts as the cathode. The electrolysis breaks down the salt to chlorine gas and sodium. Minute quantities of elemental mercury and mercuric chloride escape with the spent brine. The trouble arises if this effluent is allowed into waterways without the mercury first being removed.
3. Electrical industries using mercury.
4. Combustion of fossil fuels. Even though the mercury content of these fuels is small so much is burned that the total amount of mercury released is considerable. It has been estimated that if all the mercury in coal is vaporised on burning, then United States power plants alone could be putting 150 tons of mercury into the air each year. Throughout the world about 5,000 tons are being released from fossil fuels into the atmosphere.
5. Sludge from sewage works can contain mercury. For example, some hospitals manufacture their own mercuric

chloride cleaning solutions. This eventually ends up in the drains.

Even when the pollution stops its effect might last for another 10-100 years, unless the mercury is inactiviated or physically removed in some way.

The attempts made by the Swedish National Institute of Health to arrive at a maximum allowable mercury level in fish for consumption underline both the scientific and political difficulties involved in such activities. Initially, on November 17th 1966, it was suggested that the level should be 0.5 ppm. However, in February 1967, the Institute stated officially that fish containing 1 ppm should be regarded as unfit for human consumption without giving any reason why they had doubled the maximum allowable level of mercury.

In fact, the whole manner of arriving at maximum permitted levels appears very arbitrary. In this case an official wrote that they based their figure on the 'experience' gained at Minamata. The people who were poisoned there had eaten every day fish and shellfish with a mercury content of 27-102 ppm, i.e. an average of 50 ppm. The Swedish officials then took this figure of 50 ppm and divided it by 10, giving 5 ppm. This, on the basis of 'pharmacological experience', would be the level below which poisoning would not be noticeable. By further dividing the 5 ppm by 10, a figure of 0.5-1 ppm was arrived at, at which 'one can reasonably exclude harmful influence by methyl mercury' in fish eaten daily. That is how a maximum allowable level of 1 ppm for mercury in fish was obtained. But it appears that the officials made a mistake about average mercury concentration in Minamata fish. It seems that the Japanese expressed the concentration in relation to *dry* weight, whereas the Swedes assumed they had been expressed in relation to *wet* weight. Since a fish is about 80% water the Swedish figure is out by a factor of 5. That is, the Japanese figure of 50 ppm dry weight, used as the starting point for the 'calculations', in Swedish terms should have been taken as 10 ppm wet weight, and the net result of the various divisions would have then been 0.2 ppm *wet tissue*.

When this serious error was pointed out the National Institute of National Health announced that their published toxicological evaluation was no longer effective. A new evaluation was then produced based on a report that a man had eaten 1 mg of mercury daily for a long time without being made ill. This evaluation reached the same result as their ill-fated first report, namely 1 ppm. They also added that people ought not to eat fish from fresh or coastal waters which were polluted by mercury, but just what this was meant to do was not clear since it was not legally binding and a survey had shown that the Swedish public could not as a rule distinguish between fresh water and sea fish. Fortunately, no-one in Sweden is known to have been poisoned.

There is a valuable lesson in all this. Officials and experts are human, they are subject to many pressures, political and economic as well as scientific; they make mistakes. Decisions about the maximum levels of contaminants in food and in the environment should not be left to any single group of experts who work in secret using confidential data. Once science is done in secret it is on the way to becoming non-science, for errors—which are bound to be made—will not be picked up. In Sweden the error was spotted because at the time environmental pollution by mercury was the subject of public interest and the data were worked over by different groups. One shudders to think what skeletons still remain in official 'scientific' cupboards.

Mercury Pollution in the United States and Canada

Early in 1970 mercury pollution replaced DDT pollution as the major American pollution news-story. The work of a single scientist started the mercury furore in the USA. Norvold Fimreite, a thirty-five-year-old postgraduate student at the University of Western Ontario in London, Ontario, was studying for a Ph.D. in zoology. He recalled the mercury pollution problem in his native Scandinavia and thought that it might be worth looking for similar wildlife problems in Ontario. His experimental work, feeding hawks with mercury-contaminated liver, showed that the hawks built up a level of mercury of 18 ppm, six times higher than that in the liver. It looked as though the mercury could be concentrated to higher and higher levels along a food-chain. Fimreite later examined the pickerel, a predator fish and therefore near the top of an aquatic food-chain, and likely to show up any mercury contamination. In March 1970 he found the mercury pollution he was looking for in Lake St Clair. He told the Canadian officials on 19th March and within a week they had issued a fishing ban.

Acting on Fimreite's data, on 24th March, 1970 the Canadian authorities seized 18,000 lbs of walleye fish caught in Lake St Clair. This marks the point at which mercury pollution in North America became a political issue and the massive survey of mercury pollution began. However, as early as the autumn of 1969 the provincial authorities in Alberta banned the hunting of certain game birds because they had, like the previously mentioned Swedish birds, accumulated mercury as a result of eating mercuric fungicide-dressed seed grain.

In 1969 the United States used 6,035,000 lbs of mercury. Over a quarter, 1.6 million lbs, went to mercury cell chlor-alkali producers. Large amounts were also used in electrical apparatus, paint- and instrument-making, catalysts, general laboratory use, and pulp- and paper-making.

The Federal Water Quality Administration pointed out that in 1968 the chlor-alkali producers bought 1.3 million lbs of mercury just to maintain stocks. What happened to this 1.3 million lbs is a question the FWQA would like to know the answer to.[6] One company (unnamed) claimed that though it bought 9,120 lbs of mercury a year it wasn't putting that much into the local river. Much of it, they say, was pilfered by workers!

The Federal Water Quality Administration has stepped up its mercury analysis in American rivers and lakes and with virtually every analysis, another site of mercury contamination is found. The FWQA have released a series of names and figures which indicated the scale of pollution.

On Arthur Kill River, New Jersey GAF Corporation were found to release 28 lbs of mercury a day; they promised to cut it down to $\frac{1}{4}$ lb a day. In Calvert City, Kentucky, the B. F. Goodrich Chemical Co. and the Pennwalt Corporation polluted the Tennessee River from mercury-cell caustic chlorine units.

On the Mississippi River in Louisiana the FWQA found more mercury polluters: Kaiser Aluminium and Chemical Corporation added 12.5 lbs mercury per day at Baton Rouge and

33.2 lbs per day at Gramercy. Dow Chemical Co., at Plaquemine, added 3.22 lbs per day and Wyandotte Chemicals Corporation, Geismar, 11.7 lbs per day; and the list went on growing as more and more of the 'big names' were found to be mercury polluters. As suggestions were made that the level of mercury allowed in effluents should be zero, *Chemical and Engineering News* asked 'what effects do such controls have on the economics of caustic-chlorine production?'[7]

The difficulties of the Wyandotte Chemical Company illustrate this point. On April 7th, 1970 they were told to eliminate mercury in effluent discharged into the Detroit river. Wyandotte claims that its emergency treatment facilities, a settling tank, cost $100,000 and that they reduced the outflow of mercury from 10-20 lbs to 2 lbs per day. This was not regarded as satisfactory and a court injunction was used to force them to shut their plant down. Wyandotte have agreed to halt the discharge entirely, initially via a system which recycles spent brine (partially purified) into the underground salt formation. This still contains a risk of mercury contamination of the underground deposits. They have been ordered to produce a permanent system which avoids the underground discharge.

The former Secretary of the Interior, Walter Hickel, announced that the 'Administration is developing hard evidence and will seek court action in any confirmed case of mercury pollution if corrective measures are not taken swiftly on local levels'. Hickel also stated, 'The presence of mercury in the nation's water constitutes an imminent health hazard.'

No deaths have been reported in the United States from eating fish with high mercury content. Some people may have eaten mercury-contaminated fish for several years without showing any symptoms. However, this does not mean that all is well for, like lead, mercury is reputed to be a cumulative poison. From this standpoint methyl and alkyl mercury compounds are more dangerous than metallic or inorganic mercury and the body of a person receiving a dose of such organic mercury compounds would only have excreted half the dose after 70 days. An FDA spokesman has said that if the appearance of mercury poisoning symptoms is delayed it is impossible to predict when they might appear. Minute amounts of mercury may accumulate, becoming apparent only when irreversible damage has been done.

Scientific data on heavy metal toxicity are not good enough. For example, the FDA admits to knowing little about toxicity levels for heavy metals in food. They are firm about lead in food—a zero-level is the standard because lead is a cumulative poison. As far as mercury is concerned the FDA does not allow any inter-state trade in food with a mercury level exceeding 0.5 ppm.

The solubility of mercury in water is very low and it sinks to the bottom to do its dirty work. In fact, this misled many chemists into discounting the possibility of inorganic mercury leading to a pollution problem. Mercury in water settles in the mud at the bottom of the river or lake where it can be ingested by small organisms. These are eaten by bottom-feeding fish which are eaten in turn by larger predator fish. Thus mercury is concentrated along the food-chain.

Events have moved with the speed of quicksilver after Norvold Fimreite's discovery of North American mercury pollution. The giant chemical companies who had previously remained ignorant or ignored the problem, suddenly became prepared to consider spending up to 500,000 dollars a time on control plant, to stop mercury loss. Even so, in December 1970 Professor Bruce McDuffie of the State University of New York, who had been studying the extent of mercury contamination of fish in the Susuchana River, examined—for a joke—some tinned tuna-fish in his larder. He discovered a mercury content of over 0.5 ppm in a can. Once again the curious scientist had beaten official watchdogs.

Such a level of contamination was high enough to bring about an FDA ban. The FDA were quick to confirm Professor McDuffie's findings. They withdrew from circulation and destroyed over a million cans of tuna-fish. So it was then thought that not only fresh water was polluted by mercury, but that it had penetrated into those deep areas of the sea inhabited by tuna-fish. Professor McDuffie and his team went to work examining other fish in the market-place. They found that swordfish steaks contained up to 1.3 ppm of mercury and once again informed the FDA.

Did these findings mean that mercury pollution now extended

from the fresh water to the ocean depths, or were the fish concentrating mercury that was found naturally in the ocean? The oceans of the world contain about a billion tons of mercury, with an average concentration of one part per billion. This is between one hundred and a thousand times greater than the total amount of mercury thought to have been released into the environment by man. Thus, except in localised coastal areas, human pollution could not have increased the marine mercury level by even one per cent.[8]

Somehow, perhaps by bacteria in ocean bottom sediments, mercury is converted to the methyl form. Fish are able to concentrate methyl mercury through their gills, as well as from their diet, so that their flesh contains a level several thousand times higher than that in the surrounding water. Reports that ninety-year-old specimens of tuna have been found to contain mercury levels similar to those found in modern fish seem to confirm the view that the mercury is of natural origin. However, one must still consider the contaminated fish a potential health hazard.

The discovery of mercury in their products has been a hard economic blow for fishermen, robbed of their livelihood by this contamination. A group who used to fish Lake Erie and Lake St Clair attempted to sue Dow Chemicals. The contamination of fresh water fish also hits the tourist industry very hard. The Canadian fishing ban was imposed just before the fishing season. Lake St Clair used to be fished by 200,000 anglers.

On December 23rd 1970, President Nixon signed an executive order requiring industry to meet standards that will be laid down by the Environmental Protection Agency. No firm will be permitted to discharge wastes unless it can prove that they will not significantly lower water quality. Time and experience will show whether these measures will be adequately enforced. But even if no more mercury is allowed to enter lakes and streams, the mercury pollution problem will remain for some time. In Lake Champlain little or no mercury has been dumped since 1966, yet it is still polluted enough for the New York State to have banned the sale of its fish in 1970. It is sometimes possible for mercury-containing mud to be dredged up, but this solution is only possible for small bodies of water.

Mercury will not be the last 'unexpected' pollutant. Increasing attention has to be paid to metals such as lead, zinc, cadmium, nickel, antimony, strontium, cobalt, selenium and so on. Are they building up in the environment? Already a mysterious Japanese disease, *itai itai*, has been found to be caused by cadmium and other metal pollutants. The English translation of *itai itai* is 'ouch ouch'! It is a disease which lives up to its name. It causes a rickets-like condition and the bones become so brittle that a sneeze can break ribs.

The disease occurs in the Jintsu River Basin where the river water is polluted by cadmium, zinc and lead from mining operations. The pollution apparently started at the end of World War II. The polluted water was used to irrigate the rice crops. Initially it was noticed that crop yields fell and the farmers blamed the mine drainage water but nobody associated it with the *itai itai* which afflicted them, often fatally.

A local doctor, Norborn Higano, suspected cadmium poisoning. But when he struggled to get local mining companies to stop their toxic discharges, he was accused of seeking bogus compensation. At a parliamentary committee investigating the poisoning, years later in 1967, he broke down saying, 'I am only a country doctor ... I had no power.'

Not until after Dr Kobayashi found high levels of cadmium, zinc and lead in the tissues of *itai itai* sufferers was the connection established. Since then Dr Kobayashi has found that the amount of cadmium, zinc and lead in soil, agricultural products and water is much higher than normal in the areas around smelters. The metals apparently escape into the air during the process and are washed into the water, soil and crops. He reports soil samples which contain up to 88 ppm cadmium, 5,400 ppm zinc, and 2,100 ppm lead.[9]

To reduce the pollution of the Jintsu River a lagoon was built in which the ore and metal particles can settle out of the mine drainage water before it enters the river. The incidence of *itai itai* has now fallen in the region.

Cadmium poisoning caused a national scandal in Japan in 1970. It was dramatised in tragic fashion by the case of Takako Nakamura,[10] a twenty-eight-year-old zinc refinery worker, who killed herself by leaping from a moving train. Before her suicide, she had written in her diary, 'The fear of cadmium contamination permeates my body. Pains gnaw at me. I feel I want to throw out my stomach and intestines.' Her kidneys were found to contain 22,400 ppm of cadmium. She had absorbed the poisonous metal at work. Her death caused a great deal of comment; the Prime Minister, Eisaku Sato, is said to have wept openly, and yet the Japanese industrialists still flout legislation and escape with trivial fines and suspended sentences.

Why did it take so long for the Japanese and Swedish experiences to be appreciated in America? What were the federal agencies and the giant chemical companies doing? They had enormous information services. They must have chosen to ignore the problem until forced to act. Why is everything left until the last minute? Why did the American and British food-testing surveys fail to pick up the tuna-fish contamination?

The case of mercury pollution highlights the gross inefficiency of the official 'doomwatch' and the extent to which public awareness depends upon socially responsible scientists *outside* government and industrial organisations.

10

Industry and Pollution — The Automobile Industry

It is apparent from our previous surveys that industry is responsible for much of the current pollution problem. Our survey has tended to concentrate on the technical aspects of industrial pollution, rather than on the social and economic causal factors. In the following chapters, we shall concentrate rather more on the latter. We shall examine in some detail how three specific industries have responded to the problems they created. These industries have been chosen, not because they are necessarily the biggest sinners, but because they produce commodities familiar to the ordinary person: automobiles, detergents, and holidays.

In 1899 the *Scientific American* forecast, 'The improvement in city conditions by the general adoption of the motor car can be hardly overestimated, streets clean, dustless, and odourless ... would eliminate a greater part of the nervousness, distraction and strain of modern metropolitan life.'[1] The explosive development of the automobile industry has turned this forecast into its complete opposite.

Since the turn of the century, the automobile industry has become the pivotal industry of the modern industrial states. The expansion of the motor car industry has been tremendous —in 1939, world output was just over 5,000,000 cars per year, but by 1968 well over 20,000,000 cars were being produced, as well as 5,000,000 trucks.

The car seems to have a fatal fascination for men, literally fatal; it is the modern world's greatest killer, and in terms of a single machine, its greatest polluter. It symbolises freedom and aggression in a closed society—a symbol skilfully built up by the ad-men and given such names as Tornado, Jaguar, Avenger. What is more important is that the car does not merely symbolise power, but in fact produces immense power and profits for a handful of giant corporations which possess greater financial assets than most United Nations member states. Every effort to make cars safer and less polluting, let alone to replace the automobile by a more socially useful transport system, is faced with fierce opposition from these oligopolistic corporations.

The automobile, as we saw in our chapter on city pollution, fills the air with poisonous gases such as carbon monoxide and poisonous substances such as hydrocarbons and lead, and degrades the climate with photochemical smog. Neither man nor vegetation remains healthy in the presence of these chemicals. How can such delightful, sophisticated-looking machines be so harmful? The exterior may be carefully designed for eye-appeal, but it conceals an engine which in its functioning and effects is rather crude. Three American economists, in 1962,[2] examined the technical changes that had taken place in the American automobile between 1949 and 1960. They were able to show that the majority of model changes had been in response to superficial marketing needs rather than to genuine attempts to improve the actual running efficiency of the vehicle. These model changes resulted in cars which required spare parts more often, greater repair costs, greater petrol consumption—that is, the customer got a car that cost more to buy, more to maintain, and had a shorter life. Without even considering the costs caused by increased pollution from these models, automobile changes in the late 1950s were causing cost increases of over 11,000,000,000 dollars per year, that is $2\frac{1}{2}\%$ of the United States' gross national product. The April 1963 edition of the *American Engineer* said that: 'It would be hard to imagine anything on such a large scale that seems quite so badly engineered as the American automobile. It is the classic example of what engineering should *not* be.'[3]

There are four sources of pollution in the automobile engine: the exhaust pipe, the crank-case, the carburettor and the fuel tank. There is also pollution from asbestos particles which come off the brake-linings. Over 55% of the hydrocarbon pollutants which are produced by the engine escape through the exhaust, and the rest are equally divided between evaporation loss and crank-case blow-by. The car exhaust is composed of a wide range of organic compounds. These include paraffins, olefines, aromatics and acetylenes which are collectively known as hydrocarbons. These are produced by the combustion of the petrol in the cylinder head—they are the decomposition products of petrol. Ideally petrol could be burnt to produce only carbon dioxide and water, which are harmless. But the internal combustion engine will just not work that efficiently. An American car, without any pollution controls, will, in the course of travelling 12,000 miles, emit 500 lbs of hydrocarbons. Of the billion gallons of petrol used each year by motor cars in New York, about 70 million gallons is lost in the atmosphere as waste products and unburnt hydrocarbons. Some of these hydrocarbons may well be carcinogenic. Hydrocarbons and nitrogen oxides and other exhaust pollutants form the basis of the photochemical reaction which produces smog.

Carbon monoxide is another pollutant which results from the partial combustion of petrol. Cars produce over 65,000,000 tons of it per year in America. A single car travelling at 25 miles per hour produces 0.17 lbs of carbon monoxide per mile (and this amount doubles at speeds of 10 miles per hour). The implications of this for traffic-congested cities is very important. As the number of cars in the towns increases, the amount of carbon monoxide produced is bound to go up; the increased number of cars will produce congestion and further lower the average speed, which will further increase the carbon monoxide production of each individual car.

An enormous amount of the lead contamination of the world is produced by automobiles. Unlike the other contaminants that we have discussed, lead is not a natural combustion product of petrol. It is an additive which is added to petrol to improve its anti-knock properties (it increases the octane rating of the petrol). Lead anti-knock additives were developed by General Motors in the 1920s. Those countries, such as the Soviet Union, who have never used lead in their petrol have a much slighter lead pollution problem. Many scientists are afraid that the lead from automobiles is building up to levels that could be dangerous for health. One study, for example,[4] showed that the atmospheric lead concentration in the San Diego area is increasing by 5% a year, and that the lead is coming from these petrol additives.

The concentration of lead in San Diego is a hundred times greater than on Mount Laguna, about 45 miles away. Sometimes the lead level in San Diego air reaches 80% of the air quality limit for lead that has been proposed by the American Industrial Hygiene Association. Obviously at its present rate of increase, the San Diego levels will soon exceed this safety limit if nothing is done. There is nothing unique about San Diego, of course; all cities with heavy traffic have similar or even higher levels.

Once lead gets in the air, it does not necessarily stay there— it can get into the rain, into the land, and into vegetation. It has been estimated[5] that the use of lead additives in anti-knock petrols since 1923 has spread enough lead about to contaminate the surface of the whole northern hemisphere by about ten milligrammes per square metre.

The amount of lead in rain has been measured, and concentrations as high as 300 millionths of a gramme per litre have been recorded. The maximum level of lead allowed in a litre of drinking-water by the US Public Health Service is 50 millionths of a gramme. Plants growing near busy main roads contain high levels of lead. Grass growing near major highways near Denver has been found to contain 3,000 ppm. The amount of lead found in the grass fell away with distance from the road. Vegetables fifty feet from roads have been found to contain over 115 ppm of lead.

A litre of petrol can contain about 1 gramme in various combinations of tetramethyl and tetraethyl lead; the exhaust gases of cars have been examined and it is known that a variety of lead compounds are emitted. The total amounts that are emitted are very large—200,000 tons was said to have been emitted in America in 1966 alone.

Naturally the production of these anti-knock compounds is a lucrative business. It is worth 400 million dollars per year in sales in the USA. A further 85-90 million dollars is spent on 'scavengers', special compounds that prevent the lead from being deposited within the engines and ensure that it does its damage in the atmosphere instead. Any attack on leaded petrol is an attack on powerful vested interests. Dupont and Ethyl corporations are the two largest American producers of lead additives for petrol.

The petroleum industry also has a large vested interest in leaded petrol, as much of their advertising is based on miracle mileage ingredients which include lead. Poor-quality petrol with low octane-value can be transformed into a petrol of high octane-value by lead. The usual octane level before adding is 88; this is raised to 96 by adding lead. In 1970, over 98% of American petrol was leaded.

The 1967 Clean Air Act was the first legislation to recognise the need to control lead levels in car exhaust. This triggered off a great deal of discussion as to costs of conversion to non-leaded petrol, and over what length of time the change should be made. A study made for the American Petroleum Institute in 1967, by Bonner and Moore, estimated that the oil companies would have to invest 4.2 billion dollars to convert to producing non-leaded petrol, and that this would add 2.2 cents to the price of a gallon of petrol. The controversy over leaded additives in 1970 was very revealing as to the different interests involved. The lead-additives struggle really began to mushroom at the beginning of 1970, as the result of strong rumours that President Nixon would be introducing regulations to control the use of lead in petrol. Furthermore, the automobile manufacturers also began to put pressure on the oil refineries as they came to realise that lead would interfere with the emission control devices that they would be forced to fit to their latest models. In January, 1970 the President of General Motors, E. Cole, and Henry Ford II, both went on record as recommending a change-over to non-leaded petrol. Cole suggested that the target date for conversion should be 1975, and Ford told 19 leading oil companies that Ford would modify its engines to use non-leaded petrol whenever they made such fuel available. Chrysler too stated that it was ready when the oil companies were.

So the automobile manufacturers had put the ball well and truly into the oil manufacturers' court. During February 1970, several oil companies announced that they would be happy to supply non-leaded petrol as soon as the auto-makers had the engines ready to take it! The next announcement was by General Motors, Ford and Chrysler, later in that same month, that their 1971 models would be ready to take non-leaded petrol.

The first concrete legislative proposals came in March 1970. The Californian Air Resources Board recommended that leaded petrol be phased out over a seven-year period, and that petrol should contain no more than 0.5 grammes of lead per gallon from the new year 1971. Later in the month, the US Department of Health Education and Welfare proposed that after July 1st 1971 no petrol should contain more than 0.5 grammes of lead per gallon, and that all leaded petrol should be phased out by July 1st 1974.

By April and June, a few companies were already marketing their lead-free petrol. The leading oil companies such as Standard Oil began to complain that non-leaded petrol was costing them more to make, and would therefore be sold at a higher price than leaded petrol. Therefore, to equalise the cost of petrol, in June 1970 Nixon proposed that leaded petrol be subjected to a special tax of 2.3 cents per gallon.

Leading manufacturers of lead additives, the Ethyl Corporation, mounted a two-pronged attack on the supporters of non-leaded petrol. First they said that economic studies such as the Bonner and Moore study grossly underestimated the costs of the change-over to non-leaded petrol. The Ethyl Corporation claimed that their figures showed that the price of petrol would go up nearly 4 cents a gallon and the capital costs to produce the equipment to make non-leaded petrol would be 6 billion dollars. The Ethyl Corporation also suggested that the lead contamination from their product had not been shown to lead to any decline in health, and what is more the removal of lead from petrol would in fact increase pollution because car engines would run even less efficiently.

On the other hand, many of the design, engineering and construction companies thought that the Bonner and Moore study was unduly pessimistic; we are here referring to those concerns who would build the new refining plants to make the new non-leaded petrol. The president of one such company, Universal Oil Products Co., announced that non-leaded petrol should not cost more than 1 cent extra per gallon, and that the capital investment need be no more than 2 billion dollars.

The lead industry was beginning to be left out in the cold. This conversion to non-leaded petrol was going to hit them very hard indeed since petrol additives accounted for a quarter of the world-wide demand for lead. Petrol additives provide the American lead industry with its second largest market, about 270 thousand tons per year. The Ethyl Corporation say that the introduction of non-leaded petrol would reduce its sales by 40% and profits by an even greater margin.[6]

It is also claimed that the elimination of the lead industry's petrol market could result in the laying off of 8,000-10,000 production workers. Not only the lead industry would be affected; about 85% of sodium metal consumption is used in petrol additive production. The bromine industry also will be affected, for three-quarters of American bromine production— 75 million dollars' worth—is used in the production of ethylene dibromide, the bulk of which is used as a scavenger additive in petrol. Twelve million dollars' worth of another scavenger, ethyline dichloride, is also sold to the petrol companies.[7]

The lead industry is now forced to search for new ways of introducing lead into the environment; one of the new suggestions is to make more pesticides out of organo-lead compounds, e.g. lead-impregnated paint for the bottoms of ships which would kill off the barnacles—a graphic illustration of the toxicity of their product.

As in any other struggle to clean up the environment, the attitudes of the various corporations to non-leaded petrol differed according to how they felt it would affect their products, rather than on the intrinsic merits of the case. This was neatly put by the journal, *Chemical and Engineering News*: 'In summary, the lead anti-knock makers are resisting the conversion to a leadless petroleum industry, the petroleum companies are reluctantly beginning the conversion, the auto-makers are mildly in favour of the conversion, and the design and development companies are eagerly promoting the conversion.'[8]

Control Measures for Automobile Pollution

There are two main approaches to the problem of controlling automobile emissions. The first is to alter the design of the engines and their fuel, and the second is to try to find a complete replacement for the internal combustion engine as a transport power unit.

The three approaches to improving the pollution characteristics of the internal combustion engine are 1) to fit devices that treat exhaust gases, 2) to fit devices that alter the operation of the engine, and 3) moderations of the fuel. One control device is the after-burner system. In this system the exhaust gases are subjected to further combustion in the exhaust manifold—the Dupont Company in the United States have developed such a system which they claim almost eliminates carbon monoxide and unburnt hydrocarbons. The exhaust gases enter a shielded core of a burner and are mixed with air at a temperature of 1,653 degrees F. The after-burner is combined with a recirculation device to minimise the formation of oxides of nitrogen, and also with a trap to cut out soot and other particles. The main drawback of this device is that it is very cumbersome and increases the running costs of the car. Another way of dealing with exhaust gases is to use a catalytic converter, which breaks down some of the more dangerous compounds. The trouble with this system is that the catalyst eventually gets poisoned and no longer breaks down the exhaust gases efficiently. Lead in particular has been found to foul up the catalyst, and the realisation of this was one of the main reasons why the auto-makers began to back lead-free petrol. So if this particular device is to be used, it must be used with lead-free petrol, which in turn means that the auto-makers must produce lower-compression engines suitable for lower-octane fuel.

Another control system is manifold air oxidation, in which air is injected into exhaust valves where unburnt fuel and gases should be hot enough to ignite. Unfortunately the system is said to be cumbersome and requires high temperatures at the exhaust ports which may be difficult to achieve in small engines. To sum up, there is no available system that provides a perfect answer. Consequently, many experts have cast around for alternative propulsion units; even these seem to have considerable drawbacks, but we will consider a few of them.

It is possible to convert internal combustion engines to run on natural gas. This is nothing new; during the Second World War, many Europeans facing petrol-rationing converted their cars to run on coal gas and other substitutes. There has been a resurgence of interest in such systems, because they are so much cleaner than the petrol-burning engine. Already some operators of car fleets and trucks have switched to these systems in the USA. Not only are they cleaner, but also cheaper to run. They give twenty per cent better mileage, and natural gas is no more expensive than petrol. Their main drawback is that their range is limited to only 50 miles. This makes them suitable for local delivery work, taxis and buses, but not for long distances.[9]

The alternative to the internal combustion engine most often mentioned is the electric car. The idea of an electric car is amazingly old; the first recorded electric car ran on a track in Vermont in 1834, and in 1837, a Scotsman living in Aberdeen built a more sophisticated car which, using iron and zinc sulphuric acid batteries, produced 2 horsepower. An American called Page, in 1851, developed an electric vehicle which produced 16 horsepower. And by 1903, Baher's Torpedo broke the speed record at 120 miles per hour. But the triumph of the electric car was short-lived; the internal combustion engine, though in many ways a far cruder bit of machinery, was not lumbered with the weight of heavy batteries. The electric car, except for a few specialised uses, such as milk-floats, was not to be seen again.

Today, however, the electric car is being reassessed as a possible cure for urban transport pollution problems. Gradually industry and government organisations are beginning to do more research into these vehicles. It has been predicted that the electric car could capture 12% of new car sales in America by

1980. To get around the problem of their limited range, it has been suggested that we use electric cars in the towns and keep petrol cars for long-distance work. Most of the world's leading car manufacturers are working on designs. Ford has produced Comuta which is driven by lead-sulphuric acid batteries, with a top speed of 45 m.p.h. and a maximum range of 80 miles. General Motors have produced Electrovair 2 which uses silver-zinc batteries to give a range of 80 miles and a top speed of 60 m.p.h. General Electric have produced a model with a range of 100 miles, called Delta. It would seem therefore that there is no shortage of designs, but none of them get around the limitations of cost, weight and range. The batteries can be very heavy—those in Electrovair 2 weigh 900 lbs. If you try to reduce the weight then you reduce the performance. To get around this, we must either hope for an unexpected technical breakthrough, or accept that the benefits of the electric car outweigh its disadvantages.

In terms of cost, the electric car again mixes advantage with disadvantage; the initial cost is high—the General Electrics Delta would cost 2,500 dollars—but operation and maintenance costs are only about 1 and 0.3 cents per mile respectively, whereas current petrol cars are 2-3 and 1.2 cents per mile.[10]

An outsider in the replacement stakes is the steam car. These cars are surprisingly non-polluting and economical to run compared with petrol cars. But they present certain technical difficulties and would probably require more servicing. Apart from a few enthusiasts, no one seems to be taking them very seriously.

Gas turbine engines are also being considered but their fuel consumption is high, and whilst they produce less total pollution than the petrol engine, they do produce more nitrogen oxides.

One unusual idea that is being put forward is the fly-wheel engine.[11] Energy is stored in a rapidly spinning specially shaped fly-wheel, and this energy is easily converted into torque for powering a vehicle. The system is a streamlined bar fly-wheel spun to high speed by an electric motor connected to a suitable external power source. A fly-wheeled bus was built in 1953; it had a range of half a mile before needing respinning by a two-minute contact with an energy pole. No doubt with enough research this idea could be refined, to produce not only a non-polluting vehicle but a fairly cheap one.

There can be little doubt that it is necessary for the automobile to be changed, and possibly some of these alternative power systems will offer some solutions. But the two enormous industries, the car industry and the oil industry—the most powerful companies in the economy—are not prepared to forgo their interests without a fight. On such matters their vision is narrow and arrogant; they have one concern, survival.[12]

11
The Detergents Industry

The detergents industry has an atrocious pollution record. This industry exemplifies the situation in which the finished product rather than the actual production process is the culprit. Since World War II the detergent industry has moved through two major pollution crises. The first was the foaming problem caused by non-biodegradable detergents, and the second is eutrophication caused by the phosphate 'builder' compounds of the detergents. Even more recently, the spectres of enzyme and arsenic pollution have arisen.

The detergents industry is highly competitive. Over the years the number of firms in the industry has been whittled down until the cleansing field has become a battleground between a small number of giant industrial corporations.

As a result of this competition there has been an intense pressure to innovate, particularly those types of innovations which promise to shift the existing share of the market from one company to another. The companies have pushed one 'innovation' after another on to the housewife, often with disastrous results to the environment and the health of 'sensitive' users.

The domestic use of synthetic detergents on a large scale is a post-World War II phenomenon, but the story of their actual origin goes back over seventy years.

The Development of Synthetic Detergents

At the end of the nineteenth century and in the early years of this century, German, Belgian and British scientists discovered that certain classes of compounds had detergent qualities. These compounds were long hydrocarbon chains with a solubilising group such as sulphonate at the end of the chain. As cleaning agents these new compounds had a great advantage over traditional soap.

The soapless detergents, unlike soap, did not form scum in hard water. This scum, which sticks to fabrics being washed in it, was a great problem to the textile finishing industry and they were the first group to use the new detergents.

During the First World War the Germans were short of natural fats, the raw materials for soap. They therefore made great efforts to develop synthetic substitutes for soap. Their chemists did come up with synthetic substitutes but they were not very successful, and stood no chance of ousting soap from the home. However, they did find a market in the textile industry as a wetting agent. The textile finishing industry, in particular, found that the new compounds did not leave any scum which might adhere to their fabrics, an advantage worth paying extra for. Thus a great boost was given to the search for more efficient and cheaper substitutes for soap. During the 1930s complex phosphate 'builders' were found to reduce costs and make them suitable for washing cotton as well as wool. Later, powdered fine-wash detergents like Dreft, suitable for wool and silk, were developed, then powdered or liquid dish-washing substances. Finally, after World War II the heavy-duty general-purpose detergent-powders (Surf, Daz etc.), based on alkylaryl sulphonates and complex phosphates began to to be marketed.[1]

The appearance of large masses of foam on rivers was the first indication that synthetic detergents were pollutants. They were later found to have contaminated not only surface waters but also ground waters. This latter was particularly serious in view of the fact that ground waters are often a major source of drinking water.

The foam was created when water containing detergents, which had generally entered via the waste water from sewage, was churned up in weirs. Enormous amounts of foam created a great nuisance. It was also thought that under certain conditions they might be hazardous to health by carrying harmful bacteria from the rivers on to the inhabited land. The foam also hampered the sewage treatment process, by preventing light and oxygen getting to wastes.

A great deal of worry was caused when contaminated water entered drinking-water supplies. American Congressman Henry S. Reuse, who has fought a decade-long campaign against detergent pollution, stated: 'In Lindenhurst, Long Island... residents turned to bottled water after their taps gave out a half-and-half concoction of detergent bubbles and water.'[2]

The effects of such pollution on human health were a matter of scientific controversy. Research by the detergent industry seemed to show that the compounds were non-toxic. But some scientists found that detergent-contaminated wells tended to be contaminated with bacteria, whereas detergent-free wells were likely to be bacteria-free. Whether the presence of detergents is merely a good 'indicator' of bacterial contamination or a 'cause' of bacterial growth was never resolved. Some research was published[3] which showed that some detergents supported bacterial growth. The reason why detergents contaminated water was that they were only broken down and degraded very slowly—too slowly for the normal bacterial self-cleansing activity of water to cope with. The chief culprit was a detergent compound called sodium tetra propylene benzene sulphonate (TPBS). The reason why this compound was not degraded rapidly by bacteria is thought to be because it has a branched molecular structure. Now soap, which is *biodegradable*, i.e. can be broken down by bacteria, has a straight-chained molecule. The solution to the problem was to replace the non-biodegradable branched-chained molecules by molecules with straight chains. But they had to be equally good at washing, and also non-toxic and cheap. Such compounds were termed 'soft' or 'biodegradable' as opposed to the 'hard' or 'non-biodegradable' detergents. They include, for example, Dobane J.N. sulphonate and sodium alkane sulphonate. By the early 1960s campaigns to have the 'hard' detergents replaced by 'soft' substitutes were under way in many countries. By that time most household detergents were hard. For example, in the United States roughly 75% of all household detergents contained alkylbenzene sulphonates (ABS) and in 1962 in the USA about 560 million pounds of the substance were contained in the 4 billion pounds of household detergents sold.

Britain in the early 1960s was rather unusual in that synthetic detergents held only 43% of the household detergent markets. Nevertheless, Britain's Standing Committee on Synthetic Detergents reported that the Federation of British Industries agreed to stop making products with hard alkylbenzene sulphonates by the end of 1964.

West Germany, which had suffered a great deal of pollution, was quicker off the mark than most other countries. A bill was introduced into their parliament in 1961 to outlaw hard detergents which were not 80% degradable by October 1st, 1964. As a result of this threat, German scientists produced the first soft detergents.

The detergents industry, particularly in the United States, did not welcome scrutiny from the press and from legislators. A typical comment was: 'Hard detergents have taken more blame for pollution problems than they apparently deserve. They have become a popular target for legislators. But ... the industry has laid the groundwork for a voluntary switch to soft detergents.'[4]

What was the groundwork that laid the basis for 'a voluntary switch'? In 1963 the US detergent industry spent $5 million on research 'aimed at reducing the problem of undergraded detergents'. Although $5 million sounds a great deal, it is a mere fraction of what they spent on advertising the very products which did the pollution. Secondly, before accepting the figure for research expenditure as indicative of a top-priority effort, we need to know on what research it was spent. *Chemical and Engineering News* is illuminating: 'This research takes many forms. At one extreme, it is aimed at developing more degradable surfactants. At the other extreme, it attempts to prove ... ABS is only a very minor contributor to all-over water pollution.'[5] If the industry had really wanted to, it could have stepped up its efforts to the necessary level, to have enabled the switch to soft detergents to have occurred years before it did. An article published in March 1963 showed that there were available a number of compounds which were more degradable than the compounds then in use. But they were more expensive. It added, however, 'Price may not be a factor, though, if legislation forces a switch to new materials.' Any firm operating in an economic system based on commodity production is forced to think first of all in terms of 'price', or rather 'profit', and only secondarily, and then in terms related to the primary consideration of profit, can it afford to consider the health of consumer or environment. That is why legislation or the threat of legislation is an absolutely necessary precursor to pollution control. No detergents firm was going to pioneer 'soft' detergents in its major products before it was sure that its chief rivals were going to have to follow suit. In fact in most countries, in particular the United States, the industry struggled very hard to delay legislation until such time as they had sorted out market prices of the biodegradable replacement compounds.

The *Chemical Engineering News* of March 18th, 1963 quoted the Soap and Detergent Association as saying:

> 'Legislation will not help the situation and therefore is unnecessary. In general, detergent makers believe no legislation is required, neither state nor federal, because they feel no health emergency is involved. Producers agree that eventually they must shift their raw materials to a more degradable material and they feel they will do this as fast as possible.'

In the industry's view, the possible implementation of legislation by mid-1965 would mean that 'there would probably not be enough time for costs of the new material to come down ... therefore detergent makers would end up paying a premium which would have to be passed on'.[6]

Today the sight of foam on our rivers and waterways is no

longer as common as it used to be. The problem is now very much less than it was five years ago. But the detergents industry is not yet out of the wood as far as pollution goes. For as Kenneth Mellanby noted in 1967 the detergent pollution might have gone unnoticed, but for the appearance of the foam, and 'other cases of pollution may go undetected when new substances are used...'[7] Mellanby's prediction has been confirmed.

Phosphorus and Eutrophication

Since the late 1960s there has been increasing concern in many countries about the problem of eutrophication in fresh water lakes. At the moment phosphorus wastes are a major factor in the inadvertent fertilisation of lakes. Waste detergents are a major source of this phosphate.

Dr A. F. Bartsch, head of the US national eutrophication research programme, says that the chief reasons for singling out phosphorus as the key to control of algal growth and eutrophication are:

'1. Its supply is naturally low in surface waters to the extent that nuisance conditions of algae would not ordinarily exist. Where it can be found in waters in abundance, it usually has been added by man's activities and therefore is relatively controllable. Nitrogen, on the other hand, generally is much more abundant in soils and water and is much more soluble under aerobic conditions.
'2. Phosphorus is more easily controlled than nitrogen. Blue-green algae can fix nitrogen from the atmosphere.
'3. Phosphorus will stimulate algal fixation of nitrogen.'[8]

The damaging phosphorus originates from a number of different sources. Seventy per cent of all the phosphorus entering Lake Erie and 57% of that entering Lake Ontario, for example, are from municipal and industrial wastes. 50-70% of the phosphates in these wastes are known to come from detergents. Other sources of phosphates are: human and animal wastes, food, decayed plants, agricultural run-off, soil erosion and dissolved and suspended material in rain.[9]

We have already mentioned the two developments which made the household synthetic detergent possible: surface active agents (detergents) and phosphorus 'builders'. Now both in turn have been shown to be environmental pollutants.

A typical heavy-duty detergent contains many compounds.[10] Not all these are strictly necessary to the cleaning process, e.g. the fluorescent whitener.

The phosphates form the largest single ingredient in a modern synthetic detergent. The technical term for the phosphates' base is 'phosphate base sequestering agent' and they include compounds such as sodium tripolyphosphate (STP), tetrapotassium pyrophosphate and trisodium phosphate. STP is the most commonly used agent. The function of these compounds is to soften hard water by 'tying up' the magnesium, calcium and iron ions that would otherwise interfere with the surface active agent. This improves the cleaning-power of the detergent. Most American detergents are at least 40% phosphate, and often more. The phosphates in combination with other ingredients are said 'to increase the efficiency of the surface active agent, reduce redeposition of dirt by keeping the particles in suspension, furnish the necessary alkalinity for proper cleaning, emulsify oily and greasy soils, and help reduce germ level ... (they are) ... non-toxic, safe for colours and fibres, non-corrosive and are not flammable'. Nevertheless, are they really necessary? Do we really need either to spend vast amounts of money to remove them from waste water or to put up with eutrophication of lakes just for these advantages?

The Conservation and Natural Resources sub-committee of the US House Committee on Government Operations (The Reuss sub-committee) has recommended the complete removal of phosphates from detergents by 1972 after a detailed examination of the available evidence. The sub-committee concluded its hearings with the following recommendations:

1. The manufacture and importation of phosphate-base detergents should cease by 1972.
2. The manufacturers should immediately begin to reduce the phosphate content substantially.
3. Phosphate (enzyme) presoaks should be eliminated from the market.

4. Pending complete elimination of phosphate builders, detergents should be formulated for soft or hard waters and marketed with proper instructions.
5. The Federal Water Quality Administration (FWQA) should conduct an educational campaign against the harmful effects of phosphates.
6. The Federal Trade Commission should make manufacturers list ingredients on package labels.
7. The FWQA should support more research into the development of low-phosphate or phosphate-free detergents.
8. Immediate implementation of phosphate removal schemes for sewage treatment plants.
9. The development of programmes to control agricultural phosphorus and any new uses that might be in the technological pipeline.

The Industry's Case

The detergents industry is again fighting a rearguard action, arguing that phosphate levels in current brands are at their 'minimum practical levels', that any further reduction would reduce their cleaning power, that substitutes would increase prices, and that the current designs of dishwashers will only wash effectively with present levels of phosphate.

The spokesmen for the detergents industry argue that the alternative to a phosphate-detergents ban is the utilisation and further development of techniques for removing phosphorus from waste water. Already several local governments in the United States are attempting to reduce phosphate in waste water. Detroit, for instance, is having to spend $70 million for a treatment system which will remove 80% of the phosphates.

There are several treatment systems available. For example, ferric chlorides are used to precipitate the phosphates, and polyectrolytes are used to promote filtration of the ultra-fine precipitates. Soluble phosphorus can be removed from waste water by passing through beds of activated alumina. A caustic solution is then used to strip the phosphorus from the alumina and lime is later added to remove the phosphorus from the caustic.

Another technique passes sewage through an apparatus called a 'reciprocating-flow ion exchanger'. This makes the phosphorus precipitate out as its magnesium ammonium salt which can be sold as a slow release fertiliser.

At the time of the controversy over non-biodegradable detergents, some detergents industry spokesmen preferred a 'treatment at sewage-works approach' to the replacement of the offending compounds. The cost of treatment would be very high and a figure of $9.5 billion has been quoted[11] as the cost of phosphate removal in the USA by waste-treatment plants. Dr Weinberger, of the Federal Water Pollution Control Administration, believes that removal of 90-95% of the phosphorus from municipal waste waters can be done for 5 cents or less per 1,000 gallons in a plant treating 10 million gallons per day. This is equivalent to 1 cent per person per day.[12] This sounds little enough but it totals $3,650,000 a year for a city with a population of a million. Before we discuss the ways that have been suggested to remove the *need* for phosphates, it will be illuminating to examine the approach of the detergents industry to the phosphates furore.

The pollution charges against detergent phosphates appeared to 'throw' the industry, for initially the detergent companies did not bring to light any detailed studies of their own which would at least indicate that they had considered the possibility of phosphate pollution. Therefore they were forced to search for any work by other scientists which they hoped would cast doubts on the 'phosphate eutrophication'; for instance those who have argued that nitrogen rather than phosphorus is the limiting nutrient factor for algal growth. Of course a variety of factors, such as nitrates, carbon trace metals or hormones, could influence algal growth under certain circumstances. But the industry must also explain why the massive algal blooms have coincided with the introduction and rapidly growing use of their phosphate-based synthetic detergents.

On the other hand, the industry is forced to admit that control of eutrophication is related to the control of plant nutrients entering lakes. They then pull out their next card; 'Little is known about the biological effects of a phosphate substitute.' This has to be taken seriously and doubts have already been raised about the safety of some of the proposed phosphate substitutes. But this time the change required is not for market reasons, as were the other 'innovations', for like the bio-degradable detergents that were eventually developed after similar grumbling, the purpose of change is environmental protection. What the detergent industry would like is for the whole community to pay for their incompetence by installing phosphorus-removal systems so that they could maintain their present products.

Frank H. Healey, Research and Development vice-president of Lever Brothers Co., argued publicly, as late as 1970, that there is no replacement for phosphate, but later in the same year Lee H. Bloom, another vice-president, announced that Lever Brothers would replace phosphate with nitrilotriacetate (NTA). Mr Healey also argued further that any replacement compound would require them to design and install new chemical production plant or modify existing plant. Of course it is understandable that he objects to such a change, for it will add to Lever Brothers' costs without offering hope of a market breakthrough to offset the costs.

F. A. Gilbers, vice-president of FMC, who sold $30 millions' worth of phosphates annually to the detergent industry, also saw no point in eliminating phosphate from detergents. As an 'interim' solution he said we ought to install mechanical aeration equipment to increase effluent oxygen contents and keep algae in solution.[13]

Alternatives for Phosphates

By 1970 the American detergents industry had begun reluctantly to move towards replacing phosphate builders. The most popular replacement compound, as far as the major corporations were concerned, was nitrilotriacetate (NTA). However, the industry's plans to shift from phosphates to NTA received a great setback at the end of the year. Following a request from the American Government the industry agreed to 'voluntarily' suspend the use of NTA 'pending further tests and review'. The basis of the Governmental request was a report from the National Institute of Environmental Health Science that 'the administration of cadmium and methyl mercury simultaneously with NTA to two species of animals, rats and mice, yielded a significant increase in embryo toxicity and congenital abnormalities in the animals studied over the results with the same dosage with metals alone'. This finding could have a significance for the environment in those areas where there is widespread reliance upon septic tanks. NTA does not degrade in the anaerobic conditions which may be found in some septic tanks. Furthermore, since mercury and other heavy metal contamination of water is widespread the Government's caution seems quite reasonable.[14]

Other research shows that the breakdown of NTA might, under certain circumstances, lead to the formation of nitrosamines. This group of chemicals is known to include cancer-causing agents. Dr Samuel S. Epstein of the Children's Cancer Research Foundation in Boston has therefore argued that in replacing phosphates by NTA 'we may well be leaping from an ecological frying pan into a toxicological fire'.[15]

Unless some unusually favourable results in further experimental tests on NTA are reported, the detergents industry will have to look elsewhere for alternatives to phosphates. Potential compounds include polyelectrolytes such as polycarboxylates, these have good sequestering properties and are reasonably cheap. Unfortunately those tested so far suffer from the major drawback of not being easily biodegraded. Biodegradability is now regarded as an essential property for compounds intended for use in detergents. Another approach being attempted is the development of surface active agents that will work in hard water without the necessity of a sequestering agent.

In their efforts to avoid the use of phosphates some manufacturers have tried manufacturing detergents with a higher alkalinity. Such detergents are more caustic and therefore less safe, particularly in the hands of children who might swallow them. The task of finding suitable alternative compounds seems to be a difficult one.

Even if there proves to be no acceptable substitute compounds for phosphates the detergents industry, in theory, has alternatives. Firstly, the amount of phosphate needed depends on the hardness of water and major reductions could be made by formulating detergents according to the nature of the water in the region where they are to be sold. This would be unpopular with the manufacturers since it would increase the complexities of manufacturing and marketing. Secondly, one could reduce the usage of the existing household detergents by increasing soap usage. The Swedish Consumers' Institute have found that pre-washing with a small quantity of synthetic detergent, followed by washing with a soap product yields results equally as good as, or better than, one stage washing with synthetic detergent alone. This procedure, they claim, allows a two-thirds reduction in the use of sodium-tripolyphosphate and is successful in hard as well as soft water. Such a solution, however, calls for a responsible attitude on the part of the consumers who would be required to alter their traditional washing procedures.

Enzyme Detergents

If the detergents industry has conceded the phosphate battle, it is now faced with yet more struggles. Ever since the mid-1960s, when the industry introduced the 'biological' or enzyme-based detergent, critical voices have been raised. On the one hand for being unnecessary, any cleansing improvement being marginal, and on the other hand because they are potentially or actually harmful to the user. Such compounds have been accused of causing allergic respiratory conditions or skin reactions.

We have already discussed the damage caused to workers manufacturing the enzyme detergents. In January, 1970, the *Journal of Allergy* carried an editorial signed by a number of leading American allergy specialists expressing concern about the lack of knowledge of the long-term effects of enzymic detergent usage on the public. They wrote:

> 'we are dealing with a disease which, while it appears to have an allergic basis, remains unclear in its pathogenesis and long term consequences. As physicians and immunologists concerned with diseases of hyper-sensitivity, it appears reasonable to expect that consumers will become sensitized in their normal home use of enzyme-containing detergents ... What we understand of diseases of this nature suggests that exposure to very low levels may sensitize, we cannot estimate a lower limit which will be safe for all individuals.'[16]

Procter and Gamble, the giant detergents company, claimed, '... the experience of millions of women in Europe for ten years and in the US for several years verifies (our) scientific results proving the product's mildness to the skin and absence of allergic reactions.' Procter & Gamble tested enzyme detergents on 11,000 women for 'long periods' without skin or hay-fever-type reactions. They also claim that the consumer is exposed to only 1/50,000 of the safe level in manufacturing plants. But Procter and Gamble sell an awful lot of biological detergents, 1.6 billion normal-size boxes of biological powders per year in the USA.[17] Can they say definitely that no housewives' skin has been affected by it? The enzyme detergents also have high arsenic levels, which are discussed later in this chapter.

The fact is that all detergents are bad for the skin. This is even admitted by the detergents industry in their advertisements for their *soap* products, which they sell on the mildness to the skin.

J. I. Rodale sums up the situation. 'The chemical industry makes money on detergents in two ways. First you buy detergents which you use in highly concentrated doses, destroying the skin on your hands. Then you buy a hand lotion, or cream to counteract this poisoning, also from the chemical industry.' How can a protein dissolving enzyme on top of the normal irritant action of detergents be 'mild' to the skin?

The detergent industry seems to be extremely complacent. The journal *Chemical and Engineering News* referred to the awful damage wreaked on workers' lungs as 'past *difficulties* factory workers had in handling the enzymes in concentrated form'. These 'past difficulties' resulted in terrible pain, suffering and possibly permanent damage to the affected workers. The answer to these problems was 'new dedusting techniques, encapsulation, prilling and other means'. Once again the detergent industry brings out an unnecessary product, creating a biological problem, then shows its technological brilliance and solves it after serious damage has been done. Then it pats itself on the back by claiming it did it without the necessity for coercive legislation.

The president of Monsanto, Edward J. Bock, said in 1968 that enzymes 'represent the greatest advance in detergents in more than twenty years'.[18] One might ask, advance for whom?

The magazine *Consumer Reports* (January 1969) found that overnight soaking in regular detergents cleaned stains made by egg-yolk, chocolate, spaghetti, sauce, grass, coffee, and tea as effectively as their new biological counterparts. Even for blood stains the enzyme products were only 'slightly more effective'. However, in the year of their launching enzyme detergents and presoaks already represented a $25-30 million market in the USA and one which pundits thought would become a major segment of the $1.5 billion US detergent market.

Arsenic in Detergents

Arsenic pollution is the latest question about detergents to have been raised at the time of writing. Dr Ernest E. Angino of the Kansas State Geological Survey has found 1 to 70 parts per million of arsenic in several common presoaks and household detergents. He also found that the Kansas River contained 2-8 parts per billion of arsenic, which is close to the 10 ppb recommended maximum allowable concentration recommended by the US Public Health Service for drinking water.

He is naturally concerned: 'We are dealing with a possible pollution to our environment that is considerably more insidious and potentially far more dangerous than any reported to date.' The waste waters of these detergent products can easily enter the water system and pollute it. He believes that current arsenic levels in detergents are 'a potential health hazard to people using them constantly'.[19]

The US Public Health Service gives tolerances of 10 ppb (recommended) and 50 ppm (mandatory) of arsenic in drinking water. Angino calculated that when the arsenic-containing detergent was used as directed on the packet—and most people use more—the arsenic concentration of the washing water greatly exceeds the maximum recommended by the US Public Health Service for drinking water. Of course people don't usually drink their washing water; but, says Angino, 'the danger clearly exists that arsenic can be absorbed through unbroken skin'. Other health hazards are the creation of skin rashes and other allergic skin reactions and the inhibition of proper healing of wounds. Nobody really knows what the long-term effects of such contact with arsenic will be. What we do know is that arsenic is a cumulative poison which builds up slowly in the body, and long-term arsenosis may not be detectable for 2-6 years or longer.

Does any arsenic remain on clothing after washing? If it does one would be constantly exposed to arsenic. This might be most serious for babies, who often have broken skin where their nappies chafe them. As long ago as 1958 it was reported that arsenic became fixed in human hair after the use of arsenic-containing detergents.

Angino also showed that the treatment processes used in many sewage and effluent plants do not remove arsenic.

The American Soap and Detergent Association responded to Angino's work by claiming, 'Comparing levels in wash water to drinking-water standards is comparing apples to oranges.' They added that arsenic is chemically bound in detergent materials and can't be absorbed. They pointed out that phosphate builders in which the arsenic is found have been in detergents for more than twenty-five years and there is little reason to expect any rapid increase in arsenic levels in the environment from this source.

Presumably, therefore, they have known all about the arsenic for ages and have done the necessary research to back up their assertions. If this is the case why did they keep it to themselves? Why was it not published? Is it possible that they in fact never even gave the matter consideration until Angino published his results? Their record of biological negligence is a long one. As Mr C. G. Bueltman of the Soap and Detergents Association said in 1962[20] of their experiences with hard detergents; 'It is admitted that total consequences, and more particularly susceptibility of biological decomposition, were not anticipated. In fact, so far as biodegradability goes, hindsight shows that this was universally overlooked as a factor requiring consideration.'

The detergent industry's vast army of technologists and scientists had to produce a commodity for profitable sale, so we had hard detergents that foam, phosphate builders that cause eutrophication, enzymes that cripple workers and irritate housewives, and detergents loaded with arsenic.

12

The Tourist Industry

It is easy to forget that we may pollute whilst we play. For this reason it is worth examining the relationship between the leisure industry and pollution. There is no single leisure industry, it is a conglomerate. A whole range of industries base themselves on modern man's needs and demands for leisure and pleasure. In a society in which work for most of the population becomes ever more boring or stressful, people naturally seek diversions. Leisure is that vitally necessary break from the iron discipline of work.

As far as our investigation of pollution is concerned we need to concentrate in the main on outdoor leisure pursuits and the manner in which they are fulfilled. Whilst it is true that most leisure time is still spent in or around the home, the amount of time spent in outdoor recreation in the countryside is growing.

Hunting, shooting and fishing are traditional parts of the rural scene, But the number of people involved was in the past generally small and did not have a widespread damaging impact on the countryside. Today more and more city-dwellers flow to the countryside, and their activities continually increase in scale and kind. Thus their impact has been correspondingly greater.

In Europe almost every sphere of recreational activity, particularly motoring, water sports, and camping, has shown a spectacular growth. The number of people involved will probably have tripled by the end of the century.

In the USA, per-capita use of outdoor recreational facilities in the countryside has been increasing at a rate of about 10% per annum over the last twenty years, and it is thought that it will continue to grow.

This growth is associated with five major social mechanisms. First: a series of demographic changes, the overall growth in population, and a population increase amongst the young and the old who have most time for recreation. Second: the growth in income amongst large sections of the population in industrial countries means that there is money available to be spent on leisure. Third: a slow trend towards a shortening of the working week, and most important, a trend towards having the weekends free and longer vacations. Fourth: improvements in transport, particularly the private car and cheap air flights, have given rise to a population far more conscious of its mobility. Finally, people are becoming far more sensitive to the quality of the environment in which they live. A series of factors have contributed to this increased sensitivity. The growth of the urban population itself leads to a greater awareness of crowding or sometimes of the sheer boredom of the suburban way of life. A higher level of education, and the rise of psychological tensions in the cities, are other factors.

The room for further growth in leisure activities can be gauged from the fact that in some European countries only a minority as yet take an annual holiday away from home. In Italy only 30% of the population do, in France 45%, and in Britain 60%. In Sweden, which has the highest standard of living in Europe, the figure reaches 78%.[1]

The tourist industry exists to sell the world to this growing market. The size of the tourist industry is colossal. In many countries, particularly relatively unindustrialised countries, it is the major industry. The economic viability of such countries hinges very much on their attraction to tourists. It is the largest source of foreign exchange in Spain and even in Britain the tourist industry is the country's fourth largest export industry and biggest dollar earner. It employs five million people and has an annual turnover of £1,000m. The international movement of people engendered by the tourist industry is staggering. In 1967 over 64 million Europeans travelled to and from 13 West European countries. Tourism has become very big business.

Unless the tourist industry is very carefully controlled and planned the increased pollution engendered by so many pleasure-seeking people will destroy its very basis.

Obvious forms of pollution which are intensified by outdoor recreation are human sewage and litter. Many of the most popular tourist resorts have small permanent populations that are swollen several times over by tourists in the holiday season. Their sewage disposal facilities are often insufficient to cope with the influx. In fact, many coastal resorts have no sewage disposal treatment at all, they simply run the sewers into the sea. It is likely that such a method of disposal causes no real damage in the case of small populations. But in the holiday season it can lead to destruction of amenity. What were formerly fine, clean stretches of beach become terribly fouled. The wonderful beaches of the Mediterranean have been particularly polluted by sewage. The story is told of the visitor gazing in recent years at such a sullied beach and exclaiming: 'Now I know why every hotel here has its own swimming-pool.' Since the Mediterranean is practically tideless, the sewage situation is far worse there than it would be in tidal waters, such as the Atlantic. Dirt tends to remain on Mediterranean beaches for long periods, unless somebody takes the trouble to clean it up.

Visitors naturally object to such situations, and those countries which depend on the availability of fine beaches to attract the tourists are beginning to realise that they should become interested in improving sanitation. In the summer of 1970 the mayor of Kifo, on the Gulf of Corinth, was fired on the spot for allowing the town's sewage to foul the beach. The Greek government instituted a crash programme of beach-cleaning.[2]

All around the Mediterranean it is the same story. On the

Spanish island of Ibiza, the explosive development of the tourist industry brought about a sewage pollution crisis in 1969. The Spanish authorities had to spend £250,000 to tidy up the mess.

In Italy, throughout the summer of 1970, disease scares kept people away from formerly popular beaches in the Genoa and Lazio regions. In July 1970, a Genoese magistrate attempted to ban swimming from a ten-mile length of beach on the Ligurian coast because of the risk to health. The authorities did their damnedest to pacify the public. Dr Salvatore Custerei, president of the Italian Municipal Medical Association, said that although 'the water along the coast is dirty, polluted by solid and liquid refuse, by oil escaping from ships etc.... these things are likely to be of very real danger only to the fauna of the sea... The chief dangers to bathers are nasty little irritating skin infections and other very minor ailments which will cause them no undue harm. Certainly the present scare is unjustified.'

In Malta, the sewage system built over a hundred years ago is said to be literally 'bursting at its seams'. In August 1970, things were so bad that the Maltese Public Works Department had to employ a special boat equipped with a pump to deal with 'sewage slicks'. The Maltese are now planning to build a new sewage composting plant to deal with the situation.

Many resorts fear that tourists will start selecting their holidays on the basis of possession of clean beaches, i.e. adequate sewage treatment facilities. Consumer organisations in Britain have toyed with the idea of producing recommended lists of resorts according to this criterion. The response of the resorts was to threaten legal action if any of them were black-listed. The answer to such problems might be to tax the tourist industry to provide the cash for necessary sewage treatment facilities to be built in all tourist centres.

Another great problem is litter; not only is it unsightly, it can often be dangerous and damaging to holiday-makers, domestic and wild life. Broken glass on beaches leads to cuts, sometimes severe enough to spoil a holiday. Waste materials such as broken glass and polythene can be eaten by animals. The result may be fatal to the animal or costly to the farmer in veterinary fees. The tendency of the packaging industry to produce more and more plastic wrapping means that much of the litter is not biodegradable; if left, it will stay there for years. Degradable plastics may replace existing plastics in packaging. However, we do not know yet whether such compounds may pose a new pollution threat themselves, for the compounds into which the plastic degrades might have residual poisonous effects.

To get to the countryside people need to be transported. The cars, 'planes and boats all add to the pollution burden. More and more cars are being built; the number of cars in OECD countries rose from 21 million in 1960 to over 49 million in 1967. In Britain alone there were 13.9 million motor vehicles in 1968; if present trends are allowed to continue, this will rise to 24 million in 1980, and 33 million in the year 2000.[3] A major factor in the growth of car ownership is the convenience the car affords in allowing people to escape into the countryside. Peak leisure traffic-flows on summer Sundays can be far greater than weekday peaks for commuting and business traffic. The building of bigger and better motorways can only accelerate the process and encourage more car ownership.

The development of the day-trip tourist-trade will do likewise. A Council of Europe report notes: 'Individual entrepreneurs have found that substantial profits can accrue to those with imagination and expertise by, for example, opening up old houses, castles and game parks.' One British historic house makes an annual profit of $450,000.[4] Most of these new developments base themselves on road transport, and on popular days they serve to channel and concentrate vast numbers of cars, bringing noise, fumes, congestion and frustration into the countryside—the very things which visitors seek to escape.

The potentially disastrous effects of building motorway access to remote areas has been seen in the English Lake District, an area possessing a unique but fragile beauty. However, a reaction is setting in in the English Peak District National Park, the motor car has already been banned in the Goyt Valley at weekends.

The motor car is not the only machine of the leisure seekers that pollutes the countryside. Pleasure boats are becoming a growing source of water pollution. Wastes from a single boat are of little consequence but oil, sewage and rubbish from many boats contribute significantly to pollution. In some areas the beauty of solitude is shattered by the noise of outboard motors. The repeated passage of fast boats has often led to erosion of banks. The number of boats is growing far faster than the area of convenient inland waters in which to sail them.

In the United States there are over 8 million pleasure craft on inland lakes and streams and this total is growing at the rate of 250,000 per year. It is becoming the sport of millions; in 1969 40 million Americans went out at least twice in such boats. Very strict legislation and planning is required if a future crisis is to be avoided. The US Water Quality Improvement Act has established standards of performance for boat sanitation devices so as to reduce the discharge of inadequately treated sewage from boats.

The international tourist trade is based on air transport. More people are travelling on 'planes for non-business reasons than business reasons. In 1969 there were 5.2 million non-business passenger journeys and 3.4 million business passenger journeys from London airports. The gap between tourist and business air trips is growing. From London airports in 1980, 24.8 million passenger journeys will be tourist and only 10.6 million on business. On present trends, tourist air traffic in Britain will grow forty-fold by the year 2006 (to 276 million passenger journeys) and business flights thirteen-fold (to 74 million passenger journeys).[5] Thus more and more of the air traffic noise is going to be created by the requirements of the tourist industry, and more and more of the countryside will be gobbled up by airports catering for the tourist trade. Are we not creating a situation in which people will be leaving on holiday air trips to escape the noise created by people leaving on holiday air trips? Already people living in the vicinity of airports are fighting to get night flights stopped so that they can get a decent night's sleep.

One of the attractions of going on holiday for many people is 'getting away from it all'. Often this will take the form of going from a densely populated urban area to a less densely populated rural area. In many countries people driven by economic necessity into the cities seek to escape its pressures by buying 'second houses' in the countryside. The increase in the numbers of such second houses has been very marked in the last decade. It is an uneven development, being more marked in some countries than others. For instance, in Britain less than 1% of the population have a second house, whereas in Norway 30% do. In some countries, particularly in Scandinavia, this development has created a special tourist problem.

Ownership of a second home is a characteristic of the wealthier classes, and their desire to escape from the city may have damaging economic effects on the already under-privileged rural inhabitants. In some countries there is evidence of dramatic rises in property values. The housing market becomes so inflated by the demands of the upper income groups that local residents have difficulty in finding homes. Already in some parts of Southern France there are more second than first homes. Often second home development has been allowed to develop in a haphazard and unsightly fashion, ruining the countryside whose beauty attracted people there in the first place.

No place is safe from exploitation, today nowhere is remote. The use of light aircraft, helicopters and hovercraft will open up more and more remote areas. The use of such areas for reservoirs and nuclear power stations, itself a function of their remoteness, will further speed up this process. Very careful legislation and controls will have to be developed to protect such areas. Already the Norwegians, for example, have proposed a Mountain Plan Act to prohibit the construction of second homes and other buildings above the tree-line. They also intend to have neither roads nor other means of easy access to their national parks.

Remote areas and wilderness should belong to the world, to mankind as a whole. But because of the division of the world into national states, and because of its uneven economic development, it is likely that, under present conditions, they will be exploited in a short-sighted fashion. The solution to the problem of controlled use of wilderness lies not only within the ambit of the national state, but also with international agreement and international subsidy. Many such undeveloped regions are in poor countries, and if an area is to maintain a level of 'underdevelopment' for aesthetic and recreational reasons, it should not have to suffer economically.

Some ecosystems are particularly sensitive to increased use by holiday makers. Coastal sand-dunes are very popular recreation places, providing shelter and seclusion. However, they are one of the least resistant environments to such use, and are often used, not just for walking, loving and general messing about, but also for motor-cycle and beach-buggy races. When such activities occur on a large scale and the covering vegetation is destroyed then the sand-dunes become more mobile. Sand blows inland on to farmland, and exposes crops and moorland to harmful salt winds. This has occurred on the north-west coasts of England, Holland and Denmark. In some areas the dunes form the only effective coastal defence and their collapse, after loss of the vegetation cover which holds them together, has resulted in extensive flooding of farmland. In parts of Holland the authorities now forbid access to the dune system and access to the beaches is by way of a limited number of board walks.

The purpose of this chapter has been to show what happens when the tourist industry is allowed to grow in an unplanned, uncontrolled fashion. All over the world there are national parks, and numerous organisations staffed by dedicated individuals, all seeking to prevent the total destruction of beauty in the pursuit of pleasure by the masses. Unfortunately the money invested by the state is trivial compared to the magnitude of the problem. No amount of staring at the Mona Lisa can wear it away, but a beauty spot can literally be worn away by feet, hidden by crowds and traffic, shattered by noise, and buried beneath litter.

This need not happen if there is effective planning, education and legislation, backed of course by adequate amounts of cash. Good management techniques can be designed which will reduce the damaging effects of tourist activity; less harmful transport systems could be developed. To a certain extent the impact of the continued growth of tourism can be eased by the creation of new recreational areas. This could be done by rehabilitating derelict and under-used land; extending access to the countryside by multiple use of forest lands and reservoirs; and freer access to giant private estates.

Nevertheless, even if these improvements were made, tourism is a phenomenon which, at its present rate of growth, must eventually lead to the degradation of all unspoilt areas.

13
Military Technique and Pollution

'There is something about preparing for destruction that causes men to be more careless in spending money than they would be if they were building for construction purposes. Why that is I do not know.'
—*Senator R. B. Russell.*

The armaments industry is in the pest control business. A pest, say the specialists, is a 'species whose existence conflicts with people's profit, convenience or welfare'. The definition is easily extended to the 'enemy' so we need not be surprised if there is a carry-over in technique from one to the other. The most obvious difference between the two as far as technique goes is that the armaments industry, having put more cash and research into production, has a better chance of wiping out *homo sapiens* than the pest-control industry has of eliminating the boll weevil.

War has always had drastic environmental effects. The Thirty Years' War in the seventeenth century laid waste much of Germany for decades. But modern military techniques have created a new scale of environmental problem; they can damage even when not being used in war, and threaten those whom they mean to defend. However, before we enumerate the major pollution dangers which emerge from the armaments establishment it is necessary to examine how it has come to hold such a pivotal position within modern industrial society.

Industrial society has been characterised by two major types of war. Firstly, war by industrialised nations against non-industrial nations who became incorporated into 'empires' as colonies or forced to accept humiliating and one-sided trading treaties. Secondly, wars between industrialised nations struggling for the right to control a greater share of world markets. Consequently a large armaments industry has long been a characteristic feature of the industrial society.

However, since World War II the armaments industry has made another quantum leap. It has vastly increased in capital size, and strengthened its links with the state machine. At the same time the technology of its products changed qualitatively as well as quantitatively. Industrial society entered the age of the 'military industrial complex' and the inter-continental ballistic missile.

In 1962 the world spent £43,000 million on 'defence'. This figure is almost equal to the total annual revenue of the Third World. Seven countries spent 85% of this figure. They were Britain, the USA, the USSR, France, West Germany, China and Canada. For the last two decades, military expenditure has in these countries stabilised at a high relative level. In the USA expenditure has averaged around 10% of GNP and in Britain around 6.5%.

At the heart of the modern imperialist state are the giant corporations, the largest of which have capital assets greater than those of many nation states. Most of these corporations have developed into what have been called 'multi-national corporations'. That is, they no longer merely produce in and export from their home territory but come to rely more and more on the export of capital and the setting up of foreign subsidiaries. As home markets, particularly the United States home market, become relatively saturated, the major opportunities for expansion come from foreign markets. There is one big snag with such operations, the danger of local revolutions. If the multi-national corporations are to be able to operate wherever they want and on their own terms they are forced to seek some form of security. Thus the United States government in particular finds itself saddled with the task of providing military aid to hold back revolution. Since it is now impossible for underdeveloped countries to industrialise independently they either become counter-revolutionary and seek some form of comprador relationship with the metropolitan states or they start nationalising the property of the multi-national corporations. The latter development occurred in Cuba, and in the light of the Cuban affair, the USA and her chief allies cannot afford to allow any revolutions.

The resulting 'permanent arms economy' has gained support from other quarters. Some economists believe that we cannot survive without it, that to disarm would mean economic disaster. The economist Sumner Slichter has said of the Cold War: '(it) increases the demand for goods, helps sustain a high level of employment, accelerates technical progress and helps to raise the standard of living...'[1]

The Bomb

Apart from their actual use in war, atomic weapons pose a pollution threat to man today and in the future. Their deadly radiation can be released during manufacture or testing or by an accident to a 'delivery vehicle'. In the chapter on global pollution, radiation pollution will be more fully dealt with; at this stage I wish to stress only the manner in which the military contribute to radiation pollution.

The atomic bomb at Hiroshima caused immediate physical damage and casualties on an unprecedented scale. Sixty thousand buildings were destroyed within a three-mile radius of 'ground zero'. Somewhere between 64,000 and 240,000 people were killed (nobody really knows how many, these figures are the highest and lowest estimates).

The tragic survivors of Hiroshima and Nagasaki have been

the subjects of medical studies to discover just what long-term genetic damage was caused.

The largest study of the long-term effects has been made by the Atomic Bomb Casualty Commission (ABCC). The Commission has been conducting its epidemiological studies since 1947. It is financed, in the main, by the American Atomic Energy Commission and the research is supervised by the United States National Academy of Sciences. Most of the cash, some $47 million, and leading personnel are American. A recent review[2] summarised ABCC's main findings:

> 'No unknown new diseases have been discovered, despite popular fears that a wholly new "A-bomb disease" might afflict the survivors. But the incidence of some known diseases was clearly increased by exposure to the bomb. The incidence of leukaemia was abnormally high, reaching a peak about 1951 and declining thereafter. There was also an increase in thyroid cancer and probably other forms of cancer as well, though some of the findings are in dispute. Those who were exposed *in utero* (i.e. whilst still unborn) showed a marked increase in microcephaly (i.e. reduced brain size) and mental retardation, as well as in infant and foetal mortality ... many of the survivors of all ages were also found to have complex chromosomal abnormalities, but these abnormalities do not seem to have demonstrably damaged the health of the subjects studied. Still, the final results are not yet in, and much of the population under study is just entering the age at which further cancers might be expected to show up.'

The attitude of the US National Academy of Sciences to their findings was that they were 'grim beyond doubt, yet not so grim as many had feared'.

The ABCC findings have been criticised by Japanese scientists for distorting and suppressing data so as to minimise the horrors of atomic war; others have claimed that the controls used by the ABCC were false. In order to assess whether a population which has had a unique experience has been changed as a result of the experience, one compares it with a 'control' population. That is, one which is similar in all respects apart from the fact that it missed the unique experience. The 'unique experience' under test was exposure to radiation from atomic weapons. The critics say the ABCC's control groups were bound to be full of people who, even though not at Hiroshima or Nagasaki, had been exposed to atomic fallout from the bombs. Thus, since the ABCC compared two groups of exposed people, it is to be expected that few differences would be found. The ABCC claim to have guarded against this possibility when choosing their controls. But since everybody has now been exposed to fall-out, the Japanese probably more than most, the critics may have made an important point.

When large-scale nuclear weapons testing began in the 1950s the authorities claimed that there was no great danger. Their general line was that the consequences would be merely a slight increase in background radiation. In fact, some of the radio-nucleotides released had unforeseen properties; for instance, strontium 90 has chemical properties similar to calcium, and can be incorporated in the bone-structure of growing children, when absorbed from food. Whereas natural, or 'background' radiation is external to the human body, strontium 90 entered the food-chains. Fallout from atmospheric tests was deposited on vegetation, eaten by cattle, and concentrated in milk. Thus children since the 1950s have been absorbing a previously unknown isotope into their bodies. It is not distributed evenly, but is selectively deposited in the bones. It can do great damage for it is in the bone-marrow that the blood-cells are created. As the strontium 90 decays, it emits high-energy radiation, which can cause mutations in the developing blood-cells—leading possibly to leukaemia and bone-cancer.

Another of these fission products was iodine 131. This reaches man by the same food-chain: fall-out—grass—cow—milk—man. As with strontium 90, the radio-nuclides are not spread evenly, but are concentrated in a certain part of the body—this time it is the thyroid gland, leading to possible thyroid cancer, affecting the growth of the child. Because of the relatively short half-life[3] of iodine 131, 8 days as opposed to 28 years for strontium 90, some experts assumed, despite evidence to the contrary, that it was relatively harmless. It is true that iodine 131, falling on soil, decays before it can enter plants and the human food-chain. What these 'experts' overlooked was, however, that some of this fission product must fall directly on the leaves, where it can be absorbed by the plant before it has decayed. The result is that it is now known that children in some parts of the USA received radiation doses far in excess of those received from background radiation. These children show a marked increase of incidence of abnormal thyroid nodules.

Another unexpected discovery concerned the fission product, cesium 137. J. K. Miettinen, of the University of Helsinki, reported that in 1965 Laplanders had reached a higher radiation exposure than any other human population. Although total fallout in Lapland is half that of Helsinki, the radiation-exposure of the Laplander is 55 times greater. This is a direct result of ecological processes. Laplanders depend heavily upon reindeer meat and reindeers feed extensively on a certain lichen called reindeer moss, which concentrates cesium 137.[4] It is impossible to predict the effects of a radio-active material, or any form of pollutant, without thorough knowledge and understanding of the ecosystem.

Since the partial nuclear test ban in 1963 on atmospheric testing, the entry of such radio-nuclides into the atmosphere has greatly diminished. But it has not ceased, for China and France still conduct 'dirty' tests from time to time, and underground tests may 'leak'. Up to the end of 1970, of the 228 underground tests at the Nevada site since the signing of the partial test ban treaty, seventeen have leaked, causing radioactive contamination of the air.

Project Schooner, a 35-kiloton underground explosion in December 1968, released so much radiation that its fallout was registered in Canada and Mexico as being the highest since the end of atmospheric testing. Furthermore, since strontium 90 has a half-life of 28 years and a slow rate of fallout from the atmosphere, it will continue to kill people, though at a diminishing rate, for some time yet. Some of the other radioactive by-products of nuclear tests such as carbon 14 and cesium 137 have longer half-lives. Carbon 14 has a half-life of 5,600 years. Such compounds will continue to damage mankind for generations. Exactly how many people were and will be killed or maimed by the chronic (i.e. spread over a long time) fallout is still a matter of speculation. In a later discussion some of the estimates will be examined. The arid ground of test sites in Nevada still carries excess radioactivity which weathering and winds carry into the atmosphere. The radioactivity of airborne dust in Salt Lake City is said to increase sharply when the wind blows in from former test zones.

The very nature of nuclear weapons causes them to possess an accident-potential qualitatively different from that of conventional weapons. There are two great fears associated with accidents to atomic weapons. Firstly, that an accident might be construed as an act of aggression and trigger off a nuclear war. Second, even if the accident were not so misconstrued, it might cause terrible damage if detonated over a populous region, and even if it did not detonate there is always a possibility of radioactive contamination from damaged weapons.

The worst example of the latter type of incident occurred on 17th January 1966, off the coast of Palomares in Spain. A B-52 bomber and a KC-135 refuelling tanker collided in mid-air and four nuclear missiles were dropped. One fell without damage into a dry river bed, and two fell into farmland and were broken up, releasing radioactive material. No nuclear detonation took place; the bombs were said to be 'unarmed'. However, the radioactive contamination was so great that local crops had to be destroyed and twenty thousand sealed drums of contaminated earth taken to the USA for 'disposal'. The fourth bomb fell into deep sea-water and was not recovered until 7th April.

A fire occurred at the Rocky Flats Atomic Plant on 11th May 1969. This is near Denver, Colorado, and manufactures nuclear weapons. It is operated by the Dow Chemical Company, well-known for supplying the needs of the military with the latest creations of modern technology. This fire damaged $20 million worth of plutonium and $50 million of plant. Radioactive plutonium oxide was released, which posed a major threat to the Denver region. The Atomic Energy Commission (AEC) investigated the fire and reported that there was no hazard to the local population. Independent scientists of the Colorado Committee for Environmental Information produced evidence that indicated that 'the AEC assurances are not well founded, and that the level of contamination is yet to be fully assessed'.[5] In such a situation the public is in the hands of the official experts who tell only as much as their employers allow them to.

This is particularly so in the case of accidents to nuclear

weapons systems. In his excellent article entitled *So far, so good*, Milton Leitenberg writes, 'There are reasons for thinking that the total number of accidents involving nuclear weapons is significantly higher than the number officially announced.'[6]

Leitenberg provided a list of 33 documented accidents to US nuclear weapons systems including crashed bombers, dropped bombs, wrecked nuclear submarines and misfired missiles. He thinks that the real US total might be about 83. There are no records for the other nuclear powers, Britain, France, USSR and China, but there is no reason to suppose that they would not have had a similar proportion of 'accidents'. Leitenberg cites the story which Krushchev is supposed to have told to Nixon, then vice-president, about an occasion when a Soviet missile went off course, heading for Alaska, and had to be destroyed in mid-flight. It is thought that all the nuclear powers have evolved trigger and security systems which they believe to be fool-proof. But military experts, cocooned by secrecy from the searching criticisms of the scientific community, tend to overestimate their own powers and underestimate the powers of the forces they are playing with. As Leitenberg notes, 'The risk of accidents *per nuclear weapon deployed* may be diminishing owing to the shift to missiles and improved safety systems. On the other hand, the number of nuclear weapons deployed has been increasing rapidly and continues to do so.'

Whilst 'nuclear stalemate' may momentarily have halted the drift towards direct war between the USA and the Soviet Union, other political trends tend to highlight further environmental hazards posed by contemporary military technique. The trend towards civil war and wars of national liberation will cause a growing reliance on chemical weapons and even biological weapons, whose potential as a serious environment hazard should not be lightly dismissed.

Riot Gas

The use of poison gases by the police and military to disperse riots is growing. One of the most frequently used is CS gas. This was partly developed at Britain's Porton Down and is now sold to and used by many countries. CS gas provokes instantaneous and severe eye irritation, coughing, a burning sensation in the throat, severe chest constriction and an urge to 'get to hell out of it'. The military use of this gas is, in the opinion of some experts (e.g. Professor A. Martin, Professor of International Law at Southampton University),[7] forbidden by the Geneva Protocol on chemical and bacteriological weapons. The British government sought to avoid international embarrassment over the use of CS during the Belfast riots, by claiming that the Geneva Protocol does not cover the use of 'non-lethal' agents such as CS. On the other hand, the French government stated in December 1969 that it regards the use of CS in warfare as illegal, although it is quite prepared to use the poison against rioting students and workers.

Civil strife and rioting are on the increase and we may reasonably expect the use of CS and other control gases to increase. Certain city areas are quite likely to be periodically polluted by 'excess' and drifting riot gas. This has already happened in Belfast in Northern Ireland. The experts disagree about the toxicity of CS gas. Some claim that 'the concentration of CS gas needed to cause significant physiological damage is several orders of magnitude greater than the concentration that can be tolerated'. But it has been reported that in Belfast CS gas entered houses and certain people, particularly the old, infirm and babies, could not escape and suffered injury. This seems quite likely in view of the fact that CS is more poisonous than chlorine and that having entered the body it can be converted into cyanide. According to Dr Robert James, who has carried out research on CS at the Royal College of Surgeons in London, there is also a 'risk, admittedly slight, the sublethal concentrations of cyanide, if maintained over an adequate period of time, can cause permanent damage to the brain.'[8]

Poisonous Gases

Quite apart from their deliberate use, poisoned gases pose a number of important problems as far as the environment is concerned. First there is the danger of accidental escape during manufacture, testing, storage and transport. There is also the problem of getting rid of the gas when it is decided that its 'useful' life is over.

The most spectacular accident with a nerve gas occurred in March 1968 in the appropriately named Skull Valley, Utah. Over 6,400 sheep were killed when a test of Vx gas at the US Army Dugway Proving Ground got out of control. The gas travelled up to 45 miles, crossing two small mountain ranges on the way. Some experts believe it was a piece of luck that no loss of human life occurred, for the gas might have reached nearby Salt Lake City.

Characteristically, the US Army at first denied any responsibility. An Army spokesman claimed that they were 'definitely not responsible'. A US Congress Committee, reporting on the incident, accused the Army of 'lack of candour, deception and disregard of the public interest in its handling of such tests'.[9] After having first tried unsuccessfully to block an investigation by ranchers, the Army eventually agreed to pay the ranchers $55 a head for the loss of their sheep, which is well above their market price, in an endeavour to keep them quiet.

The congressional committee criticised the fact that no limit is imposed on the scale of such tests and that there is no requirement for the public to be informed. Representative Reuss revealed, in 1969, that the permanent Chemical Safety Committee set up 'to provide expert, independent, non-Defence Department advice on the safety of open-air testing programmes' was chaired by Dr Jake Nolan. Dr Nolan was an employee of Du Pont, who were receiving over $150 million a year from DOD contracts. Furthermore, Nolan had served as a Lieutenant-Colonel in the chemical warfare service. He was also a member of the scientific advisory group of the Test and Evaluation Commission of the US Army. No wonder Representative Reuss was driven to ask, 'with a Du Pont man as chairman how can you claim that the committee is independent?'[10]

The keeping of ex-military men on their payroll is a standard practice of companies doing business with the Department of Defence. According to Senator Proxmire, the 100 biggest military contractors employed 2,124 former high-ranking officers in 1969. The trend is to employ more ex-officers. In 1959, the 88 biggest military contractors employed 721 ex-officers, an average of 8 per company, against the current average of 21. In 1969, the top ten military contractors employed 1065 ex-officers—Lockheed alone had 210, including 22 former admirals and generals.[11]

How can anyone take US Army statements on poison gas testing at their face value? In 1969 they denied unofficial reports of tests in Hawaii. Later, when they were again caught out, an Army spokesman apologised for their 'careless handling of the truth'.[12]

The greatest problem with regard to poison gas (apart, of course, from its military use) is how to dispose of unwanted stocks. It is generally agreed that 'storing them is dangerous, burying them on land is dangerous, taking them apart and demilitarising them is dangerous, transporting them across the country and burying them at sea is dangerous'. Quite a dilemma for anyone to face. How have the US Army faced it? Their *known* record is incredible: it includes placing a nerve gas store in line with the north-south runway of Stapleton International Airport, Denver.[13] The consequences of a 'plane crashing on the site could have been catastrophic. Since being informed of this point the Army have moved their stocks elsewhere. The Army has also regularly trundled train-loads of deadly poison gas around the United States. To appreciate their faith in the transport system it is worth reminding oneself that nerve gas inhaled into the lungs acts within minutes and can be fatal in very small amounts. A victim might be saved if someone gave him a shot of atropine, artificial respiration (the mind boggles at the thought of giving the 'kiss of life' to someone with lungs full of nerve gas), and sometimes additional treatment for convulsions.[14] A large-scale disaster such as could be produced by a derailment in a built-up area would require an elaborate medical organisation. However, most of the big powers have poison gas stocks which from time to time have to be disposed of. The American Government costed the various possible disposal methods for gas from its Rocky Mountain Arsenal.

Method of disposal	*Cost ($ Millions)*
Sea burial	3.9
Land burial	11
Burning	14
Chemical reaction	17

The sea burial figures have been criticised by Dr Steven L.

Teitlebaum as being too low because they 'do not cover the very high cost of medical mobilisation to take care of a possible accident en route'.[15] If the official figure is correct, should we assume that until the highly-publicised rail trip in the summer of 1970 when nerve gas-carrying rockets entombed in concrete were carried across several states adequate medical teams were not at hand?

The background story to the now famous dumping of these nerve gas rockets off the Bahamas, published in *Science*, reads like a black comedy. 'In 1967 and 1968 the Army, acting in secrecy, had dumped more than 21,000 M-55 rockets, each armed with an explosive charge and 10.8 pounds of GB liquid nerve gas, off the New Jersey Coast. Fearing that the rockets were defective and that the gas might begin leaking out, the Army first embedded them inside steel vaults or "coffins".'[16]

In the spring of 1969, the Army had intended to dump 'on the quiet' 26,000 tons of unwanted gas weapons in the sea. The operation was named 'CHASE'—the acronym for 'Cut Holes and Sink 'Em'. However, word leaked out and the Pentagon got cold feet about CHASE. They agreed that a National Academy of Sciences Committee should advise them about how to dispose of the weapons. This committee concluded that most of the weapons could be detoxified on land but that the live rockets embedded in steel and concrete presented an apparently insoluble problem. The Academy could only recommend that the Army check with another set of experts.

The Army therefore passed the live-rockets-in-concrete disposal problem to a further committee. This committee saw only two solutions. The one which they 'vigorously recommended' was to dig a 1500-feet deep shaft, place the rockets in a cavity at the bottom and blast the lot with a 100-kiloton nuclear bomb. To do this task would have cost $3.5 million. The operation was called HARPIN. However, it required the assistance of the Atomic Energy Commission, and unfortunately for the Army the AEC didn't want anything to do with HARPIN and it was scrapped.

The only other solution was to go back to Operation CHASE. Thus in the summer of 1970 'unstable' rockets filled with a deadly gas had to be carried by rail through inhabited regions of the Southern States. They were then eventually loaded on the *Le Baron*, an old liberty ship, which was scuttled with her deadly cargo in 16,000 feet of water. Given the whole crazy set-up there seems to have been no alternative.

There is still a lot of deadly nerve gas awaiting disposal. Proposed disposal programmes in the latter half of 1970 included the destruction of 463,000 gallons of nerve gas held in 100-lb. cluster bombs, and 584,000 gallons of mustard gas. These are apparently going to be dealt with by burning and chemical neutralisation.[17]

Exactly what happens to containers of nerve gas on the seabed is not known for certain. A convenient belief is that if the containers are sunk in deep water they will do no real damage. This argument assumes, first, that any leakage of gas will be slow and will thus be neutralised over time by the great volume of water; secondly, that in the event of an implosion rapidly releasing the gas there would do no great harm because the ocean is barren beneath the sunlit level (recent studies have shown that the deep ocean contains marine life and so great damage would be done to the marine ecosystem); thirdly, that things on the ocean bed stay put and don't get shifted to some other place.

A British Ministry of Defence expert replied (18.7.69) to a series of questions, posed by the British Society for Social Responsibility in Science, about the dumping of poison gas by the British government. The official stated that they now had no stocks of poison gas, but at the end of the Second World War they had dumped 20,000 tons of German poison gas stocks in the North Sea and Atlantic. It was his belief that 'no marked effects on marine ecology are likely to occur'.[18] He gave a long list of extremely plausible arguments to back up his opinion. But within a month the *Sunday Times* reported 'Lethal mustard gas, leaking from an underwater post-war dump in the Baltic, has injured six fishermen ... it is believed that ... the containers have somehow been shifted by tides, currents or trawlers' gear into shallow waters only a few miles from holiday coasts.'[19]

Bacteriological weapons now form part of the modern arsenal, and the contamination of regions by such weapons is a kind of macabre pollution. During World War II British scientists sprayed anthrax bacteria on Gruinard, a small island off the Scottish coast. The island is now out of bounds to the public for it still harbours the deadly spores and it might be another hundred years before Gruinard is clean.

Most of the world's most dreaded diseases, e.g. anthrax, tularaemia, bubonic plague and Q-fever, have been screened as potential weapons. Research is also undertaken to increase their already great virulence as well as to develop protective vaccines.

To be effective, weapons have to be field-tested. No doubt there are several 'Gruinards'. It is known for example that the American government has field-tested diseases in at least three places: Dugway, Utah; Fort Greeley, Alaska, 80 miles from Fairbanks; Eniwetol atoll and other Pacific areas. An escape of disease organisms from such areas could lead to a large-scale epidemic. Some have expressed the view that we have been extremely lucky to have avoided such a situation so far in view of the fact that accidents have occurred. Within the USA there seems to have been one escape of organisms from a testing-ground. Traces of the disease, Venezuelan equine encephalitis, have been found in natural populations of animals near the Dugway test area. This disease of birds and rodents can be transmitted to man. It is normally only found in South America, Central America and two limited areas of the Gulf of Mexico. It is known that the US Army had been working on this disease.

Professor Steven Rose believes that the current developments of chemical and biological agents present a threat parallel to that posed by nuclear weapons. '... Just as the development of nuclear weapons indicated that physics had progressed to the point where the destruction of the planet was possible, so too, the point of no return has come with the use of chemical and biological weapons as well.'[20]

The Ecological effects of the Vietnam War

In view of the terrible human suffering inflicted on the Vietnamese people, it might seem irrelevant and irreverent to discuss the environmental damage. However, it is worth considering on several counts. The defoliation campaign was the first large-scale use of chemical warfare, whereas most knowledge of how biological and chemical agents behave is still, thankfully, theoretical. Very valuable information is coming to light about the long-term effect of defoliant sprays on the environment and on people. Further, we also learn much about the manner in which official 'experts' operate. The NLF in Vietnam is a 'people's' army. Its fighters are often indistinguishable from the peasants. They fight a guerrilla war, striking unexpectedly and then merging once again with the surroundings. Such soldiers cannot be fought by means of conventional set-piece battles. The guerrillas can only be combated, say the experts, if they can be separated from local inhabitants who feed them, and local vegetation which hides them. This was the rationale for the first large-scale military use of pesticides.

The defoliation campaign was chemical warfare waged on an unbelievably massive scale by the world's most advanced industrial nation, against a small underdeveloped non-industrial nation.

The US armed forces have flown over 30,000 poison-spraying missions over Vietnam,[21] but they have not been able to win even though they have saturated Vietnam with herbicides. 'It is to be expected that, in any future wars of this nature, more extensive use will be made of it... we must therefore... anticipate greatly expanded defoliation actions in the future.'[22]

The spraying operations seem to have begun in the early 1960s, but the idea of using herbicides in warfare is much older. The herbicides, 2,4-D and 2,4,5-T were originally developed by E. J. Kraus of Chicago University, as part of a military plan for enemy crop destruction. Apparently their military use was successfully opposed then, and during the Korean War.[23] The dubious honour of pioneering their actual use belongs to the British in Malaya. The area sprayed by herbicides for military purposes in Vietnam rose from under 18,000 acres in 1962, to a peak of over 1,700,000 acres in 1967.[24] Expenditure showed a similar escalation. In the 1969 fiscal year, the Air Force told Congress that it would spend $70.8 million on 10 million gallons of chemicals for defoliation missions in Vietnam. In 1966 the expenditure had been $12.5 millions.

A slight falling-off in spraying occurred in 1968 because of 'reassignment of equipment' during the Tet offensive. The sole limiting factor in the spraying programme is competition for equipment and men. The Department of Defense is said to have bought up the entire American production of the herbicide 2,4,5-T for 1967 and 1968. The US chemical industry was stretched to the full, and the US armed forces in Vietnam waged their chemical war on as large a scale as they could.

Spraying was not confined to forests or uninhabited regions. In 1967 over 220,000 acres of crops were sprayed, and in 1965 over a third of all spraying missions seem to have been against crops. We are talking here about deliberate spraying of crops. No doubt vast acres of crops have been destroyed as a result of drifting of sprays. For example, the College of Agriculture of the University of Saigon at Tu Duc has several times been sprayed accidentally. The result of such activity in a peasant country will be twofold, to embitter the people against the American forces, and to drive them off their land. The extent of the latter may be gauged by the fact that the population of Saigon has grown from only 250,000 to over 3,000,000 in less than ten years.

For a long time the exact composition of the defoliants used in Vietnam was kept secret and they were referred to by codenames. There are four basic types: orange, white, purple and blue.[25] There have been at least three on-the-spot studies of the ecological use of these agents.[26] All the indications are that the short- and long-term ecological effects of the defoliation campaign are very serious indeed. Forest and woodland that has been destroyed could take many decades to reach normal again. Two or more sprayings were found to kill over 50% of commercially valuable timber. This was replaced by resistant grasses, often bamboo, which are likely to hinder the growth of tree seedlings. Incidentally, where the timber industry is functioning, mills are losing up to two or three working hours a day because of damage to saw-blades caused by shrapnel in the timber. When the war is over the timber industry will obviously suffer from the effects for a long time.

Another industry hit hard by the defoliation has been the rubber industry. The Rubber Research Institute of Vietnam thinks the 'repeated defoliations are creating a threat to the very existence of rubber culture in Vietnam'.

When Orians and Pfeiffer went bird-watching in defoliated mangrove swamps they saw only one species of land bird where they would normally have found large numbers. They note that one species which does appear to be on the increase is the tiger. Apparently they have thrived so much in a situation which naturally provides them with a continuous supply of human and animal corpses that they are said now to be *attracted* by gunfire.[27] Some research indicates that 2,4-D could be toxic to fish and Orians and Pfeiffer report that the incidence of a mysterious disease of fish has increased. They add that poor people have to continue eating these fish in spite of the fact that they taste bad.

It is also known that 2,4-D can affect the metabolism of certain plant species, causing them to accumulate toxic amounts of nitrates. When eaten by animals, the nitrates are converted to nitrites which can cause a toxic compound, methaemoglobin, to form in the blood. Methaemoglobin reduces the amount of oxygen that gets to the body tissues, causing abortions in pregnant animals, or even death.

The defoliants will be spoiling the fertility of the soil in Vietnam. For example, under certain conditions, tropical soils become as hard as rock and thus useless for agriculture. This process, called laterisation, may be increasing in Vietnam. Thirty per cent of the soils in Vietnam are said to have a potential for laterisation, and the process is speeded up when soils are exposed and kept free of vegetation. The question to be resolved is just how far large-scale defoliation will open up Vietnam's soils to this destructive and disastrous process.

Orians and Pfeiffer also looked at the ecological effects of bombing. The scale of bombing in Vietnam is historically unprecedented and many parts are said to resemble a lunar landscape. Orians and Pfeiffer estimate that 848,000 craters were formed in 1967 and 2,600,000 in 1968. The craters fill with water and the resulting ponds may become centres of disease, for they are ideal breeding-places for malarial mosquitoes. Furthermore, the vast numbers of craters will interfere with efficient farming. It seems certain that the military tactics used by the US armed forces have ensured that the suffering in Vietnam will continue long after they have left.

The attitude of the military to discussions about the ecological and toxic hazards resulting from their defoliation programme is worth recording. In December 1966 the Council of the American Association for the Advancement of Science passed a resolution expressing 'its concern regarding the long-range consequences of the use of biological and chemical agents which modify the environment.[28] As a result of this resolution discussions were opened with the Department of Defense about their using defoliants in Vietnam.

John S. Foster, Jr, Director of Defense Research and Engineering, replied to the AAAS by letter stating, 'As you know, we have considered the possibility that the use of herbicides and defoliants might cause short- or long-term ecological impacts in the areas concerned. The questions of whether such impacts exist, and, if they do, whether they are detrimental or advantageous, have not yet been answered definitely, even though these chemicals have been used commercially in large quantities for many years. Qualified scientists, both inside and outside our government, and in the governments of other nations, have judged that seriously adverse consequences will not occur. *Unless we had confidence in these judgments, we would not continue to employ these materials.*' (My italics) This statement ignored research findings on arsenic compounds and 2,4-D and 2,4,5-T which were already published in scientific literature. Also one wonders who the 'governments of the nations' were who also judged that 'serious adverse consequences will not occur'.

In 1968 the DOD published a report prepared on their behalf by the Midwest Research Institute called *Assessment of the Ecological Effects of Extensive or Repeated use of Herbicides*. Amongst its major conclusions were, 'The possibility of lethal toxicity to humans, domestic animals, or wild life ... is unlikely.' 'Long-term effects on wild life may be beneficial or detrimental.' 'Herbicides now in use in Vietnam will not persist at phytotoxic (plant killing) levels in the soil for long periods.'[29]

Members of the AAAS Board which discussed this report disagreed as to its merits. Twelve out of thirteen could not share the confidence expressed by the DOD in J. S. Foster's letter, and felt 'many questions ... remain unanswered'. Three of the Board[30] went further: 'We believe that the scientific grounds for the use of herbicidal chemical weapons in Vietnam—that is, Department of Defense confidence in the judgment that they will cause no long-term effects—are not valid ... continued use of a weapon with effects that are so poorly understood raises serious moral and political questions for the US government and for the American people.'

The label on the containers of Agent White states, 'Do not

allow material to contaminate water used for irrigation, drinking or other purposes.' One wonders why Dow go to all that trouble over Agent White, which has been sprayed over hundreds of thousands of acres. They also suggest that animals should not be grazed on treated land for two years and that some crops may be damaged for up to three years after application.

The makers of Agent Blue state that when a person has been exposed to cacodylic acid over an extended period he should receive regular medical inspection. Acute poisoning by cacodylic acid can cause headaches, vomiting, diarrhoea, dizziness, stupor, convulsions, general paralysis, and death. Symptoms can be brought on by an ounce of cacodylic acid. Remember that this compound is specially reserved for crop destruction. It is therefore quite likely to be sprayed on people.

In all the discussions so far there has been the underlying assumption that military spraying techniques are the same as civilian, i.e. one can extrapolate from normal usage of the compounds. Of course this is just not true. For example, if a pilot gets into trouble—not a rare happening in wartime—he dumps his defoliant quickly. Orians and Pfeiffer reckon that a thousand gallons would be released from a spray tank under such conditions in thirty seconds.

The final blow to the credibility of the DOD line that there were no serious dangers from defoliation came when the White House Science Adviser, Lee F. DuBridge, announced on October 29th, 1969 that the US Federal Authorities would restrict the use of 2,4,5-T on food crops because it was a danger to health. Yet over a nine-year period the DOD had sprayed about 40 million pounds of 2,4,5-T over 5 million acres of Vietnam.[31]

The DuBridge announcement was caused by a report by the Bionetics Research Laboratories which suggested that 2,4,5-T had 'probably dangerous' teratogenic (foetus-deforming) effects. The most infamous teratogen is the drug thalidomide which caused hundreds of children to be born with deformities. The Bionetics results also indicated that 2,4-D showed 'potentially dangerous' teratogenic effects. The scientific issue has been complicated by the discovery that the Bionetics studies were done with herbicides contaminated by dioxin, an abbreviation for 2,3,7,8-tetrachlorodibenzo-p-dioxin. According to an article in *Nature*, 'The dioxin compound appears to be the most poisonous chemical known, excluding certain protein toxins but not the nerve gases. In the guinea-pig, dioxin is lethal in concentrations of less than 1mg/kg, or 1 part per million, which is comparable with the toxicity of Vx nerve gas.'[32] Dioxin is a very stable compound and may work its way along food-chains and accumulate in body tissues. It is estimated that something like 100 lb of dioxin have been sprayed on Vietnam.

Dioxin has been known for some time to cause a severe skin complaint amongst workers at chemical plants. In 1965 the Dow Chemical Company modified their plant so that their 2,4,5-T would not contain more than 0.5 ppm of dioxin. The batch of 2,4,5-T used by Bionics was made by Diamond Alkali Co. and has been shown to contain 28 ppm of dioxin. Thus there were suggestions in some industrial quarters that dioxin was the villain, not the herbicide.

Dr Julius Johnson, research director for the Dow Chemical Company, is reported to have stated, 'since 1950, we have been keenly aware' that highly toxic compounds could contaminate 2,4,5-T, and that they identified dioxin in 1965. The fact that the chemical industry has known for so many years about dioxin and other chemical contamination of 2,4,5-T raises some very important questions. Did the chemical industry keep this knowledge to themselves? Certainly the question of dioxin was never raised during the defoliant debate. It was, however, Dow who brought the matter up when they sought to defend their product 2,4,5-T against the Bionetics research findings. Did they ever check whether dioxin was a teratogenic compound? It seems not. The scientific journal *Nature* commented,[33] 'although information about the manufacturing hazards seems to have been shared within the chemical industry, Dr Johnson's account... suggests that the possibility that dioxin in herbicides could cause teratogenic effects in mammals was taken seriously only in late October last year (1969), when Dr DuBridge made his announcement'.

The herbicide industry argues that the Bionetics findings will not apply to their current supplies of 2,4,5-T because the herbicide is so pure that it contains less than 0.5 ppm of dioxin. But no attention has been paid to the large number of other chemicals that might be contaminating 2,4,5-T. Professor Steven Rose has analysed a sample of the defoliant spray Orange, which is a

50-50 mixture of 2,4-D and 2,4,5-T. He is reported by the *Sunday Times* to have found it grotesquely impure, containing over seventeen different compounds.[34] Nobody can say with certainty yet whether the herbicides or their contaminants are responsible for damage to people and other life in Vietnam. But to the victim it matters little.

These herbicides have been used in Vietnam on an enormous scale and with unprecedented carelessness. Arthur W. Galston, a Yale biologist, has estimated that human beings in Vietnam could possibly ingest 50 or more milligrammes of 2,4,5-T or 2,4-D per kilogramme of body weight daily by drinking water from rain-fed cisterns and ponds exposed to aerial spraying.[35]

Galston added, 'There is a possibility that the use of herbicides in Vietnam is causing birth malformations among infants of exposed mothers ... it cannot be said that the margin of safety is adequate ... although laboratory tests do not prove that 2,4,5-T and 2,4-D are able to cause birth malformations in humans at the dose levels experienced in Vietnam, the tests do suggest the possibility.'

A team of scientists representing the American Association for the Advancement of Science (AAAS) made a detailed examination of birth records in Tay Ninh, a province that had been heavily sprayed. They found that in 1968-69 over twice the national average of still-births had occurred at the Tay Ninh Provincial Hospital, 64 per thousand compared to the national average of 31.2. The AAAS team also discovered that there had been a 'disproportionate rise' in two birth defects, pure cleft palate and spina bifida, at the Saigon Children's Hospital during 1967 and 1968. They were neither able to confirm nor deny that these effects resulted from the defoliation campaigns.[36]

The Yale embryologist Clement L. Markert believed the use of 2,4,5-T and 2,4,-D posed an 'unacceptable risk' to the people of Vietnam and added that even if the compounds were not causing obvious malformations to Vietnamese children they could lead to hidden damage such as a lessening of the brain capacity.

The reaction of the DOD to all this was fairly predictable. They would, said a spokesman, restrict the use of 2,4,5-T 'to areas remote from population'. But there aren't any really large areas of military significance in Vietnam which are remote from population.

The irony of the 'permanent arms economy' is that in the long run it fails firstly to hold back revolution permanently, as the Vietnam War shows, or secondly stave off massive economic problems. Technical advances have increased military power but at the same time reduced national security. It fails to keep up employment permanently since technological advances have moved away from a military hardware whose production keeps millions in employment. Furthermore such an arms economy creates a permanent tendency to inflation. So what are we left with? The only positive result of all this vast scientific, technical and productive effort is to push mankind further into the danger of a war. We do have the hardware, it is here amongst us. We do have the men ready willing and able to use it. The atrocities of Vietnam have shown that it is idle to imagine 'civilised man' has the 'moral strength' not to use these weapons. Finally we do have the tendency to an ever deepening political and economic crisis.

14
Global Pollution

Having poisoned our own national backyards, we are beginning to poison the planet as a whole. Pollution has ceased to be a local or even a national problem. Man now faces global pollution, affecting all the countries of the world, and even those parts of the earth which are not national territory, the oceans.

Pollution-induced changes in the atmosphere may be altering the global climate and global contamination by persistent toxic materials such as radioactive substances and DDT is increasing.

In the following two chapters, we shall look at the present and future environmental problems of the sea, where there is an obvious need for international control, and a much more rational use of its resources.

Pollution and Global Climate

The climate of the earth is not constant; we know from geological and historical records that major fluctuations have occurred. There have been several ice-ages in the northern hemisphere, the last being ten thousand years ago. Since then, permanent ice has been confined to Greenland, the Arctic Ocean, a few northern islands and high mountains.

There are also climatic cycles within the main cycle. Since the last ice-age there have been alternating cold and warm periods. We know, for example, that there was a warm period between A.D. 800 and A.D. 1000, when England had flourishing vineyards. Then there was a mini ice-age from 1650 to 1840. Superimposed on this rhythm is a shorter-term rhythm. In the fifty years between 1890 and 1940, the average surface temperature of the earth rose by over 1° F. This was enough to reduce the size of many famous European glaciers. However, in the years after 1940 the average temperatures began to fall again, and by 1970 there had been a drop of 0.5° F in the surface temperature of the globe. Even this very mild temperature change may have somewhat increased the northern ice coverage, and also increased rainfall in previously arid parts of India and East Africa.

In view of the great climatic changes brought about by apparently small fluctuations in the mean global temperature, many scientists are wondering whether human activities may not produce similar changes.

One theory says that an increase in the global temperature is likely to melt the ice-caps and flood the major cities of the world. Another says that human activities are more likely to produce another ice-age, which will freeze the northern industrial societies out of existence.

Because the experts cannot agree and produce conflicting pictures, we should not therefore assume that there is a lot of fuss about nothing. On the contrary, it is even more worrying, for it means that human society is engaged in an enormous geophysical experiment with little or no idea of its implications.

There are a large number of ways in which society could be effecting changes in the climate. A recent American government report lists seven:[1]

1. Increase in atmospheric carbon dioxide, as a result of burning fossil fuels.
2. The reduction of the transparency of the atmosphere by the presence of very small particles called aerosols.
3. Another source of reduction of transparency—larger dust particles resulting from soil erosion and industrial pollution.
4. High-flying jets may be altering the properties of the stratosphere.
5. Heat from the burning of fossil and nuclear fuels may be warming up the atmosphere.
6. We may be altering the albedo, the ability of the earth's surface to reflect much of the sun's radiation back into space, by changing the nature of the surface with urban sprawls, agriculture and de-forestation.
7. Oil spills may be producing films on the sea which could alter the rate at which heat is transferred between the ocean and the atmosphere.

These changes are not pushing the climate all in one direction —some will lead to an increase, others to a decrease in temperature.

The most infamous phenomenon is the changing amount of carbon dioxide (CO_2) in the atmosphere. All the fuels we use, save nuclear, contain carbon compounds manufactured by plants. The carbon in coal used to be present in the air as CO_2 but was transformed by photosynthesis in plants which died and were transformed into coal by geological processes. In an earlier period, the CO_2 concentration in the air was forty thousand times as great as it is today. About 25% of this CO_2 was converted by plants into organic compounds, and some of this was transformed by the compression of organic sediments into the fuels that we use today. The rest of the original CO_2 was combined with magnesium and silicon that had been freed by weathering from rocks, and was precipitated on the sea floor as limestone or dolomite.

Until recent historical time, the CO_2 concentration in the atmosphere was in equilibrium between rates of weathering and photosynthesis, decay and volcanic production. However, the development of fossil fuel power has upset this equilibrium. The rate at which CO_2 is released by burning fossil fuels is a hundred times greater than the rate of release of calcium and silicon. As a result there is more CO_2 around than there has been for a long, long time.

Part of this increased carbon dioxide will be absorbed by the ocean and part by vegetation. What remains could significantly affect the environment. Although CO_2 is nearly transparent for the visible light of the sun, it has the property of absorbing infra-red radiation. A lot of the light that strikes the earth is reflected back as infra-red radiation, and the carbon dioxide in the atmosphere can trap this heat much as the glass in a greenhouse does.

It is estimated that if the current rate of increase in the burning of fossil fuels continues, by the year 2000 there will be 25% more CO_2 in the atmosphere than there was in the nineteenth century. Such an increase could raise the average global temperatures by 0.6-4 degrees centigrade, depending on the behaviour of the atmospheric water vapour content. If we did get such a temperature increase, this might well lead in turn to an increase in the amount of cloud, which would then shield the earth from the sun, and cause a drop in temperature. A 10% increase in cloud cover would cause a temperature drop of 7.5 degrees centigrade. The last ice-age was caused by a drop of 7-9 degrees centigrade.

Another process which could increase the cloud cover of the globe is the production of vast amounts of dust particles by industry, and by blow-off from agricultural land. There seems to be evidence that the amount of atmospheric dust is increasing in many parts of the world; there has been an increase of 55% over Washington, and 88% over Switzerland. This increase could reduce the transparency of the atmosphere, cutting off sunlight from the surface and cooling it down. Furthermore, dust particles can act as condensation nuclei for clouds, a further cause of increased cloud coverage.

There is already some evidence that increased dust levels cause local climatic changes. LaPorte, Indiana, is thirty miles downwind from the steelworks of Gary and South Chicago. Between 1951-65, LaPorte had 31% more rain, 38% more thunderstorms, and 24.5% more days of hail, than nearby communities. It was found that the amount of rain falling on LaPorte depended upon the amount of steel produced.

Excessive dustiness may have quite the opposite effect in regions where there is relatively little moisture in the atmosphere. In this case, too many very small droplets are formed which remain in the atmosphere as cloud instead of falling as rain. In the sugar-producing regions of Queensland, Australia, the cane leaf is burnt off before harvesting, creating huge palls of smoke. The smoke palls are said to reduce rainfall in the area by 25%.[2]

It has also been said that exhaust from cars may affect clouds. A typical car exhaust contains as many as 10,000 sub-microscopic particles of lead per cc. Iodine in the atmosphere, produced from the ocean and from forest fires, could be reacting with the lead to form lead iodide. Lead iodide is the substance that rainmakers use to 'seed' clouds. This process may be affecting the rainfall pattern in certain areas, which would account for the fact that in the north-eastern states of the USA the amount of light misty rain, or 'Scotch mist', and finely divided, slow-falling, 'dusty snow' has been increasing in recent years.[3]

Aircraft themselves may well be producing more cloud. The vapour trails caused by jet aircraft sometimes expand to form an extensive layer of cirrus clouds. Dr Bryson claims that high-flying jets have already increased the cirrus cloud cover between North America and Europe by 5-10%. He also thinks that the introduction of several hundred SSTs, such as the Concorde, might increase the cloud cover by up to 100%, which could result in severe climatic changes. But not all scientists in the field agree with Dr Bryson's calculations; they argue that the cirrus clouds produced by the jet aircraft would have appeared anyway, and that the jets merely hastened their appearance. At the moment no one is very sure just what effect increased coverage of cirrus clouds would have on the temperature. It has been suggested that the cirrus clouds resulting from aircraft could affect the circulations of storm systems around the globe.[4]

Despite all the contradictory views several things are clear from the current discussion of the effects of pollution on the global climate: firstly their potentially catastrophic effects, and secondly the obvious requirement for far more research so that we can understand the effects of our actions, the natural instability of climatic conditions, and the interaction between them. It is quite likely that knowledge will advance rapidly in this field, but there is no guarantee that concrete governmental action will ensue.

Global Contamination by Organochlorine Insecticides

Organochlorine insecticides are, as we have already mentioned, probably more widespread than any other synthetic chemicals, and have become one of the world's most serious pollution problems. DDT residues are found in soils that have never been treated with insecticides, in the rain, in the air of remote regions, in the wildlife and snow of the Antarctic, and probably in all animals and humans generally.

Once released into the environment, these compounds can be moved by physical factors, such as water and air, and by biological factors. Agriculture is the major, but not the only, source of organochlorine insecticide pollution. These compounds are also prevalent in rivers receiving effluents from the chemical, textile, and clothing industries. Firms manufacturing insecticides may from time to time release some of the insecticides into rivers. The textile and clothing industry use a great deal of insecticide for insect-proofing materials, and may be even less aware than agriculturists of their dangers. Once in the rivers, the compounds soon reach the sea and enter marine food webs.

Air currents are believed to be responsible for much of the global distribution of organochlorine insecticides. They get into the air by volatilisation from the soil, and by spraying. Charles Wurster claims that once in the air, transport is so rapid that they may circle the globe in a few weeks.[5]

A major factor in the polluting effect of organochlorine compounds is their persistence—they are not easily broken down, and can have a half-life of 10-15 years. Within a food-chain, such compounds increase in concentration with increasing trophic level, the highest concentration being the carnivores at the highest trophic level. Biological concentration results from the fact that each organism eats a great many smaller organisms from the next trophic level—small vegetarian fish eat plants, and in turn are eaten by larger fish, which may be eaten in their turn by seals, or by man. The predator digests, metabolises, and excretes the flesh of its prey to a greater extent than the insecticide residues. Much of the insecticide residue is thus retained and tends to accumulate. The increase in concentration of the contaminant between the prey and the predator may be a mere doubling, or occasionally a thousandfold increase, according to rates of intake and excretion.

Clear Lake, California, is the classic example of biological concentration. An organochlorine insecticide DDD was added to the lake at concentrations of no more than 20 parts per billion over a three-year period up to 1957. Biologists studying concentrations in wildlife at Clear Lake, found that: the plankton in the lake concentrated the DDD to a level 265 times greater than the amount to be found in the water; the plankton the fish fed on had residues 500 times greater; and carnivorous fish, and fish-eating birds, contained residues over 80,000 times greater than the original concentration. It is this mechanism that has resulted in some carnivorous organisms, particularly birds, carrying residues more than a million times greater than their local environment. By 1960 the local Western Grebe colony had decreased from 1,000 to only 30 pairs; DDD residue levels of 500-1,500 parts per million were found in their body fat. The colony had been virtually wiped out by the insecticide. In 1967, ten years after the last DDD application, the Grebe population had only risen to 165 pairs.[6] At least 14 bird species are known to have suffered similar setbacks in many regions of the world.

It would, however, be a mistake to assume that all the harmful effects of DDT and related compounds are restricted to the top trophic levels of the food webs. Charles Wurster has discovered that a minute concentration of a few parts per billion of DDT in water can decrease the growth and photosynthesis rate of certain marine plankton. Because of the fundamental basic position of these plants in marine ecosystems Wurster believes that the effects that he observed 'may be ecologically more important than the obvious direct mortality of larger organisms that is so often reported'.[7] No one can say whether the changes which Wurster observed in the laboratory are occurring in the oceans. We just do not know enough about plankton population fluctuations over time.

Manwell and Baker believe that the failure to predict that persistent pesticides would build up in food webs was a major technological blunder.[8] They point out that since the 1930s, most of the ecology textbooks discussed food webs, and that textbooks on comparative and general physiology contained copious discussions of active transport, and examples of the accumulation of substances from the environment—for example, that certain ascidians (small marine animals) could concentrate vanadium in their tissues at over a billion times its concentration in the surrounding water. It seems likely that the oversight resulted from a combination of factors, such as commercial pressures, and the fragmentation of science itself.

Radioactive Pollution

We have already discussed the dangers of the military uses of radioactivity; we will now discuss it from the point of view of peaceful uses. A nuclear reactor is not a bomb. Nuclear weapons have been specially designed to produce as big a bang as possible, whereas a nuclear reactor is based on the control of the fission process. The energy is conducted away to do its useful work, whilst the dangerous fission products remain behind—or should. In theory, it is possible to produce a perfectly safe reactor, but in practice they are not perfect and can become dangerous in certain circumstances. For instance, at Windscale in Northern England there was an incident in 1957 when large parts of the surrounding area became contaminated, and milk from local farms became unusable. There have been quite a number of other dangerous incidents.

The design of nuclear power stations is improving and their efficiency increasing, and as this form of power is moving out of the experimental stage, so pressure is increasing all over the world for more commercial stations to be built. It is at this stage of development that many people are beginning to fear the dangers which may result from large-scale industrial operations. Previously, stations were small in scale, carefully monitored, and staffed by very well-trained supervisors, but with the coming of mass production, can we ensure that standards will be maintained?

One of the more disturbing features of recent developments in the industry has been the considerable number of component failures in reactors. In 1970, there were so many failures in American reactors that the licensing authorities were forced to propose a tightening-up of design criteria. There was a special tendency for pressure vessels to develop faults. In the 640 MW Oyster Creek nuclear power station in the United States, 123 out of 137 welds to tubes connected to the pressure vessel were cracked. A similar thing happened at the Tarapur station in India, which the United States were building for the Indian government. In Italy, at Trino, another American-designed station developed pressure vessel failure. A nuclear power plant at Chooz in Belgium had to be closed down because the bolts that held the nuclear furnace together had disintegrated.

The legislation governing nuclear power reactors has, however, been from the start far more stringent than in any other industry. We have shown how in other industries it has taken a series of catastrophes to get stricter pollution control measures. This industry was different in that it started with a catastrophe —a catastrophe on a scale that ensured that nobody could remain ignorant of the effects of radioactivity—the bombing of Hiroshima and Nagasaki. The other important difference is that it was not created by private industry but by government organisations, staffed by scientists more concerned with technical perfection than with balance-sheets and profits.

The potential health-hazards of nuclear power stations can be reduced by careful siting. It has been generally accepted that you do not have people living in the immediate vicinity of the power station—this is called the 'exclusion area'; it is completely under the control of the station, so in the event of mishap, evacuation can be completed rapidly. Beyond the 'exclusion area', the surrounding districts should be of low population-density—again, so as to enable easy evacuation if necessary. There are also regulations assessing how far the power station must be from any large centre of population.

According to Sax, for a 630 MW nuclear power station the safe distance for a centre of population would be just under 10 miles, for a 200 MW station, $4\frac{1}{2}$ miles, for a 50 MW station, just under 2 miles.[9] Sax goes on to say that 'in practice, reactor sites are frequently chosen by the potential licencee for economic and other considerations, rather than purely from the criteria set forth in 10-CFR-100' (a Federal Register).

Obviously, as the number of nuclear plants grows, in some countries it is going to prove difficult to stick to these regulations, and companies are going to try to get round them. The early stations were built in very isolated places, but now we read quite often about cases of townspeople protesting at proposals to build stations near by. At Stourport, England, local inhabitants successfully prevented the Central Electricity Generating Board from siting an advanced gas-cooled reactor at a spot only 1½ miles from the town centre.

The disposal of nuclear waste products is a very difficult problem. These products will range from hot water to highly radioactive materials. Radioactive material cannot be disposed of in a conventional manner because it cannot be biologically broken down and can in some cases remain highly dangerous for hundreds of years. In the USA, over 67 million gallons of such material were produced in 1960. The usual disposal method is to store it in a liquid form in specially cooled steel and concrete tanks designed for a long life, placed well away from population centres. But on at least one occasion these tanks were sited on a faulty area in the USA liable to earthquakes. Consequently it has been argued that this can only be a temporary solution and that the radioactive waste should be fixed in solid material such as glass which could be sealed in stainless steel cans for storage. Another approach has been to inject it into underground formations, which has led in some places to earth tremors. The final standby is to seal it in concrete and dump it into the sea. A group of West European countries, including Britain, in 1968 dropped nearly 36 thousand such containers, weighing 11,000 tons, into deep water. The supporters of this approach claim that the long-term effects and risks are small. But there is always the possibility that the containers will develop faults and release radioactive material into the sea—this has happened often enough with nerve gas containers in the Baltic. It is in any case a very short-sighted view as man will have to make increasing use of the sea, if he is to survive in the future. We may complain today about the mess left by early industrialists—slag-heaps and jerry-building—that we have to clear up today; yet the nuclear power industry is leaving material that is not only unsightly, but lethal, that will need watching for hundreds of years. We are saddling future generations with problems. This, more than any other factor, is why large-scale nuclear development should not be allowed at its present stage of waste-disposal technology.

The least radioactive material is disposed of by dilution by water or air. One of the dangers of releasing even low-level radioactive waste into rivers or the sea is the danger of radioactive materials being taken up by living organisms and concentrated as they pass up the food-chains. At Windscale, the discharge of radioactive waste into the sea has to be very carefully controlled. It is known, for example, that edible seaweed is capable of absorbing radioactive iodine. Therefore the emission limits have to take into consideration the fact that a small percentage of the British population in fact eat this seaweed as a delicacy.

Atmospheric discharges are also very closely controlled, and discharges from chimneys are subject to a special filtering and cleaning process which removes, for example, iodine 131 and ruthenium. However, radioactive noble gases, xenon and krypton, pass through the filters. Since these are noble gases and inert they do not combine into chemical compounds, but remain as circulating gases in the atmosphere. Krypton 85, because of its relatively long half-life of 10.76 years, poses in particular a tricky problem. Krypton is removed from the atmosphere and concentrated for various industrial purposes, and by the end of the century, the presence of krypton 85 in the atmosphere could make these processes dangerous. Methods for removing this substance are now being seriously studied, but so far without success.

A number of commercial groups are co-operating with the US Atomic Energy Commission in an attempt to use atomic explosions for civil engineering tasks, for example, blasting canals and reservoirs.

Atomic explosions are much faster than normal excavation, but obviously there will be tremendous pollution problems, on a scale that was associated with nuclear testing. Radio isotopes such as strontium 90, cesium 137, and iodine 131 would be released into the atmosphere on a large scale. But in spite of this, the AEC has gone ahead with such proposals as Project Plowshare. Such projects would make nonsense of all present safety restrictions, and some of Plowshare's supporters, among them Edward Teller, 'father of the hydrogen bomb', are currently waging a campaign to have these restrictions weakened.

The most serious proposals so far contemplated for the commercial use of nuclear explosives concern new ways of releasing underground gas and making oil production from shale fields economically feasible. The first try-out of this approach was Operation Gasbuggy. Notice this playful, nursery-like name, indicating no doubt the equally childlike lack of responsibility. In this project a 26-kiloton bomb was exploded 4,000 feet below the ground in New Mexico. The explosion produced a 'chimney' into which gas from the surrounding rocks would seep. The technicians knew of course that the first lot of gas that they got out would be radioactive, so they flared it, releasing radioactive krypton 85 and radioactive hydrogen (tritium). The scheme misfired in two ways; first of all the gas was not of good quality, containing too much carbon dioxide, and secondly, its radioactivity failed to drop below the allowed upper limits, even after 300 million cubic feet of gas had been removed. The tritium that was discharged is potentially one of the most dangerous isotopes because, being hydrogen, it may form radioactive water which could be taken up by plants, passed into food-chains, and reach human beings by way of meat and milk. It is calculated that at its present rate of release from all sources, the maximum permitted safety-level will be exceeded in the twenty-first century.

The next Plowshare experiment was Project Rulison, with a 40-kiloton bomb 8000 feet below ground. On that occasion a traditional 'dirty' fission bomb was used in an attempt to avoid the tritium problem. Such a bomb would produce strontium 90, cesium 137, krypton 85, *et cetera*, which in theory should remain underground. At the time of writing the organisers have not flared the gas because of local objections from people naturally frightened of possible radiation leaks.

One of the big unknowns is just how much damage exposure to continual low dosages of radiation causes. Man has always been subjected to natural, background radiation which may be responsible for certain spontaneous mutations; what we are now considering is an extra dosage from the nuclear power industry. Is it going to increase the cancer rate and genetic aberrations in man and other living things? Professor Sternglass has claimed that radioactive fallout has caused the death of 400,000 infants in the United States since the early 1950s. He suggested that the ingestion of strontium 90 makes babies more susceptible to normal diseases, and he noted that the infant mortality rate had declined continuously from 1935-50; after 1950 it levelled off for a number of years. This levelling coincided with the commencement of the Nevada atomic tests, and then after the atomic test ban treaty the mortality rate resumed its decline. Sternglass's results have been bitterly criticised but his fears that low dosage exposure may be more dangerous than was previously thought might prove to be correct.

There are other sources of radiation; for example, X-ray equipment, colour television, and the radioactive materials used in medicine, research and general industrial tasks. Then there are non-ionising radiations such as lasers, microwave ovens, communications equipment, and ultrasonic cleaning devices.

Every day people are X-rayed; this increases the mutation rate in the population. Children born of mothers who were given foetal X-rays during pregnancy show a 40% greater occurrence of leukaemia. Obviously X-rays are an extremely valuable diagnostic tool, but a great deal more care is needed in using them.

In 1956, the National Academy of Sciences (NAS) Committee on the Biological Effects of Atomic Radiation recommended that the contribution of all man-made radiation to the human body should not exceed ten thousand billion rem[10] per generation (thirty years). This is approximately 0.33 rem per year. They estimated that exposure from medical uses of radiation already accounted for about half this value. Thus the NAS and the Federal Radiation Council (FRC) arrived at a figure of 0.17 rem per year as the maximum permissible exposure to be allowed from the nuclear power industry. The reader may gain an idea of the magnitude of this by considering the exposure he is already receiving from natural sources of radiation. This background radiation varies from one place to another, ranging from 0.05–0.2 rems per year: in a few places people can receive about 1.5 rems per year.[11]

Two leading radiation scientists, John Gofman and Arthur Tamplin, believe that if the entire population of the USA were ever exposed to the full 0.17-rem exposure limit, from nuclear power stations, then there would be an extra 16,000 cases of cancer per year, and also a large increase in genetically determined diseases. They argue that the current regulations for nuclear power stations should therefore be made even more stringent.[12]

All the existing regulations assume that exposure is a cumulative effect which grows linearly with the build-up of the dosage; that there is no threshold below which damaging effects are absent. Damage therefore occurs in direct proportion to the radiation dose. However, this is the most pessimistic view and it may be that low doses of radiation are less damaging unit for unit than are higher doses.

The Atomic Energy Commission argue that the general public is not being exposed at the moment to any more than a small fraction of the maximum limit of permissible radiation, because existing control measures are so careful. If this is true, why all the worry about making regulations even tighter? The reason is probably because they know present standards cannot in fact be maintained under pressure by the growing industry. Nuclear power in 1969 provided only 1.4% of the total electrical power produced in the United States. But it is estimated that the total by 1980 will be 25%, and by the year 2000 it may be 60%.

The drive for profit is forcing the development of nuclear power on to an industrial scale before full knowledge of the safety factors involved has been reached. A more socially rational approach would be to operate only on a developmental scale until the safety and disposal problems have been completely sorted out. The need for such a stringent approach stems from the unique character of nuclear pollution. Not only does it kill and cripple the current generation, but it also endangers the existence of future, unconsulted generations.

15
Pollution of the Marine Environment

The seas are man's hope for the future; the last frontier, breathing-space for the land which is being over-exploited. For this reason, if for no other, any threat to this resource by pollution must be fought. The pollution pressures are mounting up; populations move to the coasts seeking the better amenities of such regions, but industries are tending to follow them, finding there labour and convenient sites. International trade is greater than ever before in human history and increasing amounts of cargo, much of it noxious, are being transported on the sea. This creates an acute threat. The sea is polluted from all sides; by waste from ships, by rivers and even by fallout from the air.

Ironically, pressures to get a cleaner environment on the land create new threats to the sea. Faced with increasing treatment costs and higher standards on the land, the industrialists and city authorities, ever on the lookout to keep down or cut costs, look to the sea as some kind of infinite bottomless pit for their wastes. It is free, once you reach it, and for the most part not protected by legislation, except in the case of oil discharges.

A number of events of the last few years stand out and symbolise the rape of the last earthly virgin—the sea. We may remember the voyage of the *Lucky Dragon*, the Japanese fishing-boat which came home covered with radioactive dust; the undersea search for a lost hydrogen bomb off the coast of Spain; the wreck of the *Torrey Canyon* and the *Santa Barbara* oil seepage. The detection of DDT, PCBs, mercury and other toxins in marine life over the whole globe emphasises the unity, the oneness of the earth and the dangers of global pollution.

Some Facts about the Sea

The oceans are about a hundred million years old and their water is an extremely complex solution of inorganic and organic components. The inorganic salts result from millions of years of erosion of the earth's surface, and from volcanic and meteoritic material; the organic matter from the remains of life of the sea and land. Today man is adding to these 'natural' components of sea-water. The water itself is thought to have been released from previously molten rock as the earth cooled down.

One of the main causes for concern about the use of the sea as a dump for 'awkward' waste materials is that, despite the rapid growth of oceanography, we still know relatively little about the oceans. Our ignorance is reflected in the fact that oceanography is not yet a single science, but a mingling of many sciences oriented towards solving problems about the sea.

The problem of what sea-water is has been a central one for oceanography for over ninety years, and yet an exact, universal description of ocean water cannot be made. Sea-water is an extremely interesting solution containing most of the naturally occurring elements. Most of them are found only in the minutest concentration; only 14 elements occur in concentrations of more than 1 part per million. Ninety-nine per cent of the dissolved constituents of sea-water are salts of chloride, sodium, magnesium sulphate, calcium and potassium. A very remarkable feature of sea-water is that the ratio of any one of these major constituents to the total of dissolved solids remains almost constant over the whole globe, even though the total amount of dissolved solids varies from one place to another. This is a result of the global effectiveness of the water-mixing and biological cycling processes of the ocean.

This leads us to a most important property of the sea as far as understanding pollution is concerned—oceanic circulation. Oceanic circulation has several causes. As water evaporates from the sea surfaces the density of the surface water increases. The denser water then sinks, causing currents and circulation. It is also (more frequently) caused by cooling and ice formation. When ice is formed in polar regions the brine left behind has a greater density since it is colder and higher in salt concentration than the original water.

Dissolved matter is moved about in the sea by mass transport, diffusion and biological transportation; by physical factors such as currents, evaporation, and sedimentation; and by geochemical factors, such as volcanic action and chemical reaction.

There is also a biomass cyclic system, a passage of elements from water through living organisms and back to the water as a result of assimilation and decomposition. Not only energy but particular compounds may travel through food webs.

Photosynthesis occurs generally only in the top 100 metres about the maximum depth penetrated by sufficient light. In this so-called euphotic zone the photosynthetic organisms utilise carbon dioxide and nutrients such as phosphates and nitrates. These floating plants or phytoplankton are the primary producers of the ocean ecosystem. Furthermore, they release vast amounts of oxygen and play a vital role in maintaining the oxygen level of the atmosphere.

Not all living creatures in the sea depend so directly on light. The ocean also contains chemosynthetic and heterotrophic bacteria. The former are able to obtain energy for manufacturing carbohydrates not from light but from a variety of inorganic substances. The heterotrophic bacteria derive their energy from breaking down dead and decaying organic matter. Such bacteria occur at all depths in the sea, even to thousands of metres below the euphotic zone.

Sea-water is no longer thought of as a simple solution, but as an extremely complex system. 'The interplay amongst the organic, inorganic, living and non-living components in seawater, and physical processes, such as ocean-mixing, turbulence, and light penetration, is extremely complex. They are all inter-related and inter-dependent.'[1]

Currents or other factors cause, in some regions, an 'upwelling' of deep water towards the surface. This is ecologically important because it brings up the deeper waters that are rich in nutrients. The nutrients enable a plankton population to be maintained, which in turn leads to larger fish populations. Although ocean bottoms are rich in nutrients they cannot be utilised by photosynthesising organisms because light does not penetrate more than 300 metres down. Currents are also the means by which highly toxic matters thought to be safely on the sea-bed for eternity could enter into food-chains utilised by man.

Marine Pollution

Marine pollution has been defined as the 'introduction by man of substances into the marine environment resulting in such deleterious effects as are harmful to living resources, hazards to human health, hindrance to marine activities, including fishing, impairment of quality for use of sea water and reduction of amenities.'[2]

Major polluting processes of the marine environment were seen as

(1) Domestic sewage
(2) Industrial waste: in approximate order of current importance.
   (a) Pesticides, which would include agricultural run-off.
   (b) Heavy metals.
   (c) Radioactive materials.
   (d) Petrochemicals.
   (e) Oils, etc.
   (f) Pulp and paper waste, including inorganic waste.
   (g) Detergents.
   (h) Heat.
   (i) Solid objects.
   (j) Dredging spoils, including mining or quarrying waste material.

In this chapter we will concentrate mainly on pollutants which have originated from land-based activities, and leave our discussion of oil pollution and undersea mining to a separate chapter.

The ocean is seen by many waste-disposal 'experts' as the perfect dump. Their views are worth quoting at length.

'The great economy inherent in the discharge of urban sewage and industrial waste into near shore waters for final disposal is apparent to all who will investigate. It is doubly apparent to those charged with the responsibility of disposing of such wastes without excessive cost to the public or menace to public health. If the ocean, or one of its arms, can be reached with a sewer outfall, within the bounds of economy, the grim spectre of an expensive complete treatment plant grows dimmer and dimmer until it fades entirely and, to the great satisfaction of those who have had to gather funds for the public budget, as well as they (you and I) who have to pay the bill, the good old ocean does the job free.

And small wonder that we look to the sea for this assistance. Its vast area and volume, its oxygen laden waters, its lack of potability or usefulness for domestic and most industrial purposes, present an unlimited and most attractive reservoir for waste assimilation.'[3]

'Ocean disposal of waste waters presents opportunities which are unique in the area of waste water treatment and disposal. Unlike most situations the availability of the ocean makes it possible not only to consider various alternative combinations of treatment and out-fall location but also, and most importantly, to consider out-fall location with respect to minimising disturbances to the ecosystem and the associated beneficial uses. That is, based on economics, where applicable, and technological and scientific evaluations, it is possible to locate an out-fall along a coast line where the *negative benefits* [my italics] are minimised while still accomplishing waste water disposal. Although such an approach is most desirable in any waste water management system, ocean disposal is one of the few instances where it can be realised.'[4]

Note the stylistic changes, e.g. the use of 'ecosystem', whilst in effect still maintaining exactly the same outlook as the previously quoted engineer. All that separates these two quotes is five years, during which it has become necessary to set up smoke-screens in an attempt to confuse conservationists. Note also the lengths to which such people go to avoid using what they would term 'emotive' words. They write of 'negative benefits', which is 'the sort of language which if it is carried to its logical extreme, will replace "dead" by ... "negatively alive"'.[5]

A British government technical committee on the disposal of solid toxic wastes claimed in 1970 that 'the amount of water in the seas and oceans of the world is almost unbelievably great and its potentialities for dilution and rendering harmless the wastes of mankind are almost infinite.'[6]

As a result of following the kind of policies advocated by such advisers, certain areas of the ocean are showing signs of severe damage. Generally such areas are still, fortunately, localised—semi-enclosed bays and harbours in heavily populated regions, for example.

An American study examined the surface waters of the Atlantic coast of the USA for suspended matters, particularly solid pollutants swept into the sea by rivers and sewers. They found that 'appreciable amounts (more than 1 milligram in a litre of sea-water sample) were restricted to an area within a few kilometres of the coastline'.[7] That is, the materials swept into the sea tend to move along the coastline rather than straight out to sea. Such findings seem to show that the deeper sea on the continental slopes is relatively free of this type of pollution, though Thor Hyerdahl's *Ra* expedition reported seeing very foul water in some places on the transatlantic voyage.

According to Jacques-Yves Cousteau, the French marine scientist, 'life in the sea has diminished by forty per cent in the last twenty years'. Cousteau does not back his statement with figures but it is the impression of a scientist who has spent a lifetime on and under the sea.

The Baltic Sea

The Baltic has been particularly hard hit by pollution of every possible type.[8] The Baltic has peculiar geographical characteristics which have intensified the pollution. It has for natural reasons shown a tendency to become stagnant from lack of oxygen. It is almost totally land-locked with only three very narrow outlets to the North Sea and its coasts house over twenty million people, often in great industrial cities such as Leningrad, Copenhagen, Stockholm, Helsinki and many others. These pour a great variety of waste matter into the sea. As a result of these conditions the concentration of oxygen, so essential to life, has continually decreased since the beginning of the century. For example, the concentration in the bottom water of the Northern Central Basin of the Baltic has decreased from slightly over 30% of saturation in 1900 to zero in 1970!

Enormous amounts of sewage are discharged into the Baltic. Most of the sewage is untreated or only partially treated. The semi-enclosed bays and archipelagos, where water exchange with the open sea is restricted, are the worst affected areas. The amount of phosphorus being discharged in sewage is increasing but the full implications of this for the sea are not yet understood.

The worst industrial pollution is that caused by the Swedish and Finnish wood processing industry. Heavy metals, phenols and cyanides, *all extremely toxic*, are amongst substances discharged into the Baltic. Though their quantities are said to be small some scientists believe that there may be long-term dangers. Often the pollution starts inland where industrial plants discharge their effluents into inland lakes and rivers. The rivers then carry polluted water into the Baltic. As a result Baltic marine organisms have shown increasing levels of mercury, DDT and PCB. Some idea of the degree of pollution can be gained from the fact that DDT and PCB levels are eight times higher in Baltic wild life than in comparable creatures from the west coast of Sweden or England. Pesticide pollution may be expected to decrease since most of the Baltic countries have now barred or restricted organochlorine compounds.

There has also been considerable oil damage on parts of the Swedish coast.

Some dumping in the open sea has occurred. In the years 1930 and 1931 Sweden dumped 7.6 million kg of raw arsenic mixed with concrete into the Bay of Bothnia.

The Baltic region was the scene of some of the most bitter fighting in World War II and this too has left its legacy of pollution. There are still mines lying on the bed of the Baltic or floating in its waters; over a hundred ships have been destroyed *since* the war by such mines. Also, at the end of the war, ammunition and poison gas were dumped into the Baltic.

A more balanced view of the effects of marine pollution can be obtained by examining the effects of various categories of waste.

Sewage Disposal

Even those countries which seriously control the inland disposal of sewage are extremely lax about putting untreated sewage into the sea. In Britain, which has, by most existing standards, relatively stringent rules about putting sewage into fresh water, local authorities are allowed to discharge untreated sewage into the sea estuaries and tidal rivers. A recent British government[9] report recommended that:

If sewage is not treated in sewage works, then the following minimum standards should be met.
1) Sewage should not reach bathing beaches.
2) The discharge site should be far enough away from the shore to render it inoffensive to people on the shore or bathing.
3) Outfalls should be sited so as not to interfere with local fisheries.

The general effects of the disposal of sewage into the sea will be, firstly, to lower the oxygen level of the receiving water; secondly, to encourage the growth of bacteria and phytoplankton. Such 'blooms' which result from an over-fertilisation of the water can be of toxic species. These may be eaten by shellfish which then harbour the toxin. Outbreaks of poisoning have been caused by eating such shellfish.

Thirdly, some bottom-living species may be eliminated in areas near outfalls. For example, species adapted for living on clean rocky bottoms may become smothered and replaced by species which prefer silty conditions.

A continuing worry is whether a danger to public health is

created by pouring untreated sewage into the sea. There have been occasional typhoid outbreaks resulting from sewage contamination of bathing beaches, but usually the carrier has quickly been identified and treated. It is generally believed that in countries with well-organised public health systems this problem has been satisfactorily controlled. But we just do not know enough about the long-term effects of ingesting small amounts of toxins from effluents, either on marine organisms or on people who might eat these creatures.

The degradation products of oil, which include carcinogens, may be accumulated in fish. Also, fish from chronically polluted waters are more prone to precancerous conditions. Some people argue that a causal link with pollution has to be proved conclusively before we do anything which might affect industry. But one must remember just how hard it was to get 'conclusive' proof even in the now well-documented cases of cigarettes and lung cancer and the ecological dangers of DDT.

Industrial Waste

Like domestic sewage, industrial waste contains large amounts of organic matter, but in addition it often contains an unlimited range of toxic chemicals. The only industrial effluent which is strictly controlled is the discharge of oil from tankers. For the rest, the rule seems to be 'do as you like'. As legislation tightens up on the land, so more and more industries prepare plans for dumping wastes in the sea by pipeline or from tankers. The sea is being currently poisoned by cyanides, mercury compounds, pesticides, polychlorinated biphenyls and heavy metals.

Many of these pollutants are dealt with at length elsewhere in the book. Of particular interest are the effluents from the petrochemical industries, sited so often on the coast. Petrochemicals dumped into the sea have tainted fish. This is an economic blow to the fishermen who catch the fish and are then unable to sell it. We are not so well off for food that we can afford to make any of it inedible. Many of the petrochemical substances that are discharged are carcinogens and even if they are not taken up by man (and that is not yet proven), they make the fish diseased. In some localities this type of pollution depletes the fishery resources because it reduces the oxygen level of the water. The problems created by petrochemical pollution do not remain at the national level, for fish and other forms of migratory life do not respect man's national boundaries. Unless there is a major change in the situation, petrochemical pollution of the oceans will increase, leading to serious pollution in the future.

Heavy metals are particularly dangerous toxic pollutants because they accumulate in living organisms. These generally come from industrial wastes which are discharged into rivers and estuaries or even dumped directly into the sea through special pipelines. Such pollution was often regarded merely as a local problem, but we now know that heavy metals are accumulated within some organisms which are able to spread their effects over a wider area.

A piece of good luck brought to light a most serious pollutant a few years ago. When scientists in the mid-1960s were analysing wildlife for DDT and other organochlorine pesticides, they also noticed the presence of strange organochlorine compounds which they could not identify. These mysterious compounds were so similar in structure to DDT that they caused interference with DDT measurements. They were found to occur most frequently and in the largest proportions in the livers, fat, and eggs of birds, especially predators such as the kestrel and sparrow-hawk, and in marine feeders, like the guillemots and kittiwakes. They were also found in fresh water fish and seals from the Scottish coast.

Soren Jensen, a Swedish scientist, solved the problem of the identity of these mysterious pollutants.[10] They were polychlorinated biphenyls (or PCB). PCB is not a single compound, but the name of a whole family of compounds with a common basic structure. They were first manufactured as long ago as 1929 by the Monsanto Chemical Company of America and have a wide range of technical uses in the electrical industry, and in the manufacture of paints and varnishes, synthetic adhesives, and lubricants.

PCBs are known to be poisonous when absorbed in accidental, acute dosages. They have caused workers to develop skin disease and liver disease. But until their fortuitous discovery in wildlife, nobody had realised that the effects of exposure to PCB were not limited only to the immediate location of use.

If the compounds had been first manufactured in 1929, how long had they been in the environment? Soren Jensen was again able to provide a plausible answer. He looked for PCB residues in eagle feathers in a museum, and only found it in specimens caught after 1944. This probably indicates that PCBs had been environmental contaminants for over twenty years before they were first noticed.

A most disturbing finding about PCBs is that, like DDT which they greatly resemble, they can have an insidious effect on the hormonal system of birds, in particular upon their calcium metabolism. It is not known whether or not they can interfere with our hormonal system. Reports of finding these chemicals have now come in from all over the northern hemisphere.[11] Over a period of two months, beginning in September 1969, at least eight thousand dead birds were washed up on the Irish Sea coasts of England, Wales, Ireland and Scotland. Some of the birds were examined for the presence of toxic chemicals. The analyses revealed concentrations of hundreds of parts per million of PCBs. But nobody knows whether the chemicals were responsible for killing the birds. The British Government Chemist said in his annual report (1970) that PCB residue levels in wildlife were rising; levels more than twice the previously known highest levels were now being found in sea birds.

Riseborough and his American colleagues have noticed their presence in a wide range of American birds, and have also noted that residues are highest in industrial areas.[12] They believe that PCBs are probably now distributed in the continental ecosystem of the United States.

Exactly how PCBs get into the environment is not fully understood. Monsanto Chemicals claim that it is impossible for PCBs to escape from their plant during manufacture. But they add that during the manufacture (by other firms) of PCB-containing paints, varnishes and coatings, factory pipes leak the chemical. It is carried in effluent into the sea, where it is accumulated and concentrated in wild life. The disturbing feature of PCB pollution is that nobody even suspected its presence. It had been accumulating in the environment for over twenty years before it was finally discovered by accident.

It seems quite likely that as we start to look for pollutants the PCBs will be found not to have been an isolated 'surprise'. Others no doubt lie in wait. The current trend towards industrial waste-dumping by pipeline into the sea is criminal madness. We just do not know enough to say with the confidence of some 'experts' currently backing such schemes that destruction of nursery areas of young fish and shellfish, and other far-reaching effects, are impossible or unlikely. Even more crazy is the fact that there is no compulsion on companies to publish, even for the scientific community, the contents of their effluents, or to remove known toxic compounds.

There are very sound biological reasons why extra-special care has to be taken with heavy metal pollution. Marine organisms accumulate metals from sea-water by factors which vary from less than 1 for sodium to more than a million for zinc.

There is also a danger that radioactive elements may be returned to man through food-chains. One of the most worrying has been radio-nuclide contamination of edible seaweed. This can be extremely important in those regions which have a taste for kelp—South Wales, for example.

Unless an organism has the ability to regulate the concentration of a particular metal in its body the concentration of that metal in the tissues will increase as the concentration in the water increases. It is hard to generalise about what the effect of such accumulation would be. But it could poison the organism, and even a slight toxic effect can have a serious effect on the ecology of the organism. If the compound is accumulated along a food-chain the consequences, as we saw in the case of mercury pollution, can be very serious indeed. Whether or not a particular polluting metal will or will not be accumulated is difficult to predict. The process is known to be complicated by such factors as the metabolism of specific organisms, its chemical state in sea-water, and the presence or absence of other metals and compounds. Similarly complicated is the removal of metals from organisms. Information is particularly sparse on the chronic effects of low concentrations of metallic pollutants.

Dredging Spoil and Mining and Quarrying Waste

Mining and dredging operations in some places lead to a heavy concentration of suspended materials, which can cause serious changes in marine life, extending sometimes up to fifty miles from the coast. In many areas the dumping of spoil is affecting spawning and nursery grounds. In special cases increased fish population might be found in an area of turbidity, but it is more often accompanied by a reduction in the number of species. The most dangerous aspect is damage to nursery grounds, where the young fish grow and develop, for this can have wide-reaching effects on fishing. Off St Austell in Cornwall a china clay mining company intends to build a pipeline to discharge clay residue about a kilometre offshore at a depth of 18m. Over the next sixty years they estimate that some 200 million tons of the residues will be dumped in an area of less than 6 square kilometres. They claim that this will not harm local fisheries or amenities, in fact they suggest it will improve them. Do they really know enough to speak with such confidence? Since the capital costs of the scheme are over £5 million pounds it could prove an expensive mistake, and therefore, once it is operating, the clay company would have a vested interest in keeping open the pipeline.

Pesticides

Many pesticides have accumulated in the marine environment, having reached the sea in a number of ways; from rainfall, down rivers, from agricultural and forest treatment run-off, sewage, marsh spraying and log-boom treatment, from accumulated spillage and from certain industrial processes like moth-proofing. The relative quantities from these sources are not known, though most will probably enter the sea via the rivers. Very few rivers are monitored for pesticide content; even in Britain, which is relatively advanced in this field, the river authorities are only just beginning to be supplied with the necessary analytical equipment. It is surprising how little is known about the effects of these substances on the marine environment, their persistence and ultimate fate. For example, even though the annual use of DDT in America has declined over the last decade, in California the amount of DDT residues in marine phytoplankton (small floating plants) has increased threefold since 1955. Since the phytoplankton form the base of the food-chains in the sea, this research seems to indicate that some delay should be expected before the bird life, which has been hard hit by DDT, begins to recover. A great deal of work remains to be done before pesticides' effects can be measured and the degree to which they pollute the sea understood.

Pesticides applied to land in Africa have been detected in the Bay of Bengal and the Caribbean Sea, having been transported by the summer monsoons and north-east trade winds respectively. Analyses of fish and birds, from the Atlantic and Pacific oceans, show that DDT is being accumulated in marine life. The Bermuda petrel is a rare sea-bird that does not have contact with any land mass treated with insecticides. Yet, according to two American biologists, it now carries a burden of DDT residues, and its breeding success has declined since 1968.[13] It could only have picked up the DDT from a marine food-chain.

The situation in the Pacific seems much the same. Many species of sea-bird and fish have now been found to be carrying DDT residues. These residues are higher in some marine fish than in fresh water fish from regions where there are heavy pesticide applications to surrounding farmland. 'The implication,' says Charles Wurster, 'is clear. The failure of salmon and trout to produce viable fry in Lake George and Lake Michigan may soon be repeated among some of the world's major marine fisheries—if it is not already occurring.'[14]

Certain types of pesticide are now used on the sea itself. Over two thousand marine species are said to be pests, damaging man or his property. In the mid-1960s, the damage was estimated to be $500 million per year; the American Navy spends over $100 million each year on control, damage repair and replacement. The range and variety of pests is enormous. Some, such as the ship-worm (Teredo), bore into piles, docks and other wood structures. Barnacles and other species with sessile life-stages grow on ship bottoms and get into the intakes of pipes for power stations. Algae and other marine life block and clog pipe intakes and filters. If, as seems likely, man's use of the sea increases, then more of its denizens will be regarded as pests. A growing market for marine pesticides will form, and the demand will be satisfied—possibly with damaging results. Certainly the experience with agricultural pest control and pollution gives no cause for confidence that such a situation would be dealt with any more wisely.

Radioactive Materials

Radioactivity is more carefully controlled than other industrial pollutants. This is as it should be, for the potential for damage to man and the environment is so much greater. Nevertheless, there is a steady stream of radio-nuclides entering the sea from atomic power stations, nuclear-powered vessels, and from the testing of atomic weapons. Some of these enter food webs and are accumulated.

The maximum permissible doses for man, and the maximum levels for radioactive discharges, are worked out using the most critical pathway by which radioactive isotopes are concentrated or transmitted to man. For example, at Windscale in northern England, where low-level radioactive wastes have been poured into the Irish Sea since 1952, the maximum level of radio-nuclides is based on the fact that certain edible seaweeds concentrate radioactive iodine. Fairly large amounts of certain seaweeds which could be contaminated are eaten by people in South Wales. In fact, seaweeds are harvested in many parts of the world, not only as a delicacy but as cattle food and sources of iodine. They also provide extracts such as agar, algin and carageenan which are used as thickeners by the food industry.

Not everybody is happy with the view that these 'low-level' discharges are innocuous. For in actual fact very little is known about the effect of these low levels of radiation on organisms.

Soviet researchers experimenting with pelagic eggs of Black Sea anchovies found that perceptible biological effects resulted from strontium 90 at levels that were close to permissible levels. In the view of G. G. Polikarpov, the maximum permissible limits have already been reached and 'further radioactive contamination of the seas and oceans is inadmissible'.[15] Already in the Black Sea the amount of radioactive matter in seaweed is nearly ten times greater than it was before radioactive pollution.

Just to emphasise the complexity of the task of determining

what are damaging levels, American workers using salmon eggs found no such damage.

Some countries dump their high-level radioactive wastes into the sea sealed in drums or concrete blocks. These are dropped into the deepest parts of the ocean. Unfortunately knowledge of deep ocean currents and sediment movements is incomplete, to say the least, and such procedures are too risky to be allowed to continue.

Heat or Thermal Pollution

This is generally associated with the power-generating industry. The increased use of coastal sites for power stations makes the sea, at least locally, susceptible to thermal pollution. There is evidence that heat in sea-water encourages the development of certain undesirable organisms, such as teredo in wood, and interferes with the migration of fish in some North American salmon runs. Such changes might have widespread repercussions. On the other hand, warm water in some cases assists survival in severe conditions and encourages the growth of certain forms of fish that might be useful.

Fish Farming[16]

The oceans provide 15% of the animal protein eaten by man and improved fishing techniques are increasing the world's fishing yields by 6% per year. Obviously if overfishing is to be avoided and such yields are to be maintained and even increased, there will have to be a greater willingness to accept national and international controls.

The culture of marine fish is likely to be increased rapidly. Already many food firms, such as Unilever, are making a serious study of the problem. At the moment the concern is to produce increased amounts of expensive 'quality' fish such as Dover sole rather than a direct effort to solve the world's shortage of protein. 'Culture' is the practice of rearing and growing young fish from captive parent stock. This has been done for centuries in fresh water fisheries, particularly in Japan, where they have developed great expertise. So far, in spite of much research, a 'commercial breakthrough' has not yet been made.

To farm fish you must have a suitable site and plenty of unpolluted water. One problem is that large-scale farms are expensive to build and the expense is only justified if they can be built close to their market. But it is rare to find the right quality of water near to the urban centres. Several schemes have been proposed which would combine power-station complexes and aquafarming schemes. Control systems have to be available to prevent fish being exposed to excessive heat, particularly in the hatcheries at spawning times. The warm water under the right conditions can be used to advantage, say British experts. They have proposed a system called 'raceway farming' which uses a limited space combined with a high throughput of water. The system could be used at the nuclear power complex at Hunterston. When operational, Hunterston will be discharging nearly 40 million gallons of water each hour, which is equivalent to one complete change of water per day in ponds six feet deep covering an area of a square mile. Using 10% of the Hunterston flow in raceways, they estimate that over 1,500 tons of fish per year could be harvested. Some idea of the scale can be given if one remembers that in 1968 the total landings of Dover sole in England and Wales weighed 1,737 tons. In such a system very great care would have to be taken to see that the cultured fish contained no higher levels of radio-nuclides than do wild fish.

Such ideas show that, given enough knowledge and the necessary planning, the combined production of power and protein is a most likely development. The system would require the delimitation of zones in which industry, agriculture and local authorities would have to meet more stringent pollution control. Pollution could therefore hold back the utilisation of this technology for a long time. Pollution is obviously more dangerous to the cultivated fish since they are enclosed and living at high densities and are therefore, unlike free fish, not able to escape from 'poor quality' water. The eggs in hatcheries are even more sensitive and there would have to be legislation giving aquafarms priority over polluting industries.

Such visions should not blind us to the possible dangers. Just as modern agriculture has poisoned the land, so might aquafarming poison the sea. The cultured fish will be susceptible to various diseases and parasites and there will be great pressure to use pesticides, e.g. algaecides and fungicides. Furthermore, the aquafarming may use intensive feeding and fertilisation and eutrophication will be a danger.

Conclusion

The activities of industrialised society have already put the sea under strain despite its enormous size. Dr E. L. Cronin of the University of Maryland believes that for the safe disposal of effluents from the East Coast of the United States some 300 billion cubic metres per year of sea are needed and he adds that this is approaching the annual exchange of all water on the Eastern continental shelf. The sea in that area is already within measurable distance of reaching its capacity to absorb wastes.[17]

The sea is going to be more and more exploited for both its mineral and organic resources. But the world cannot afford to rape the sea as it has raped the land, for increased fish catches will be a vital factor in redressing the world's protein shortages and raising the average diet standards for hundreds of millions living on inadequate diets. Pollution could jeopardise the fishing industry.

Since the Second World War fishery production has doubled, and the annual catch of sea fish is about 50 million tons. This represents about 15% of the world's animal protein consumption. This could be greatly increased.

Some experts[18] believe that by the end of the century the catch can be raised to 200-300 million tons. Such an increase is possible if present known stocks are conserved and not overfished, if already known but unused fisheries are exploited, and if new fishing-grounds are discovered. The discovery of new fishing-grounds is thought most likely;[19] the principal upwelling areas and some continental shelves seem to be very fertile, though little-known.

The Antarctic Sea is thought to contain enough fish to support yields as large as or larger than the present world fish production. But so far, because of its distance from world markets and the consequently high capital costs of the necessary fleets, only its high unit-value crop, whales, has been harvested. Already the USSR and Japan have started exploratory work; others will follow. The bitter experiences caused by overfishing of the newly-discovered marine salmon zone off Greenland show that without stringent international controls any new fishing-grounds will be overfished.

Will this exploitation be carefully planned internationally and done in conjunction with an international ecological monitoring scheme? Will there be a powerful overseeing organisation with enough teeth to halt potentially ecologically dangerous activities? Will such an organisation have the power and ability to co-ordinate the myriad of oceanic enterprises? This is what would be required if the sea is not to suffer pollution on the scale experienced by the land. In the present political situation, for such planning we would need revolutionary changes in the international structure. The sea is seen as fair game for anybody with the necessary cash and determination. Although international agreement succeeded in dividing up the North Sea continental shelf into national zones, the exploitation of the seabed poses extremely tangled legal problems. Historical experience shows that technical problems seem to be solved more rapidly than legal problems. The present legal situation seems to be that a nation can only exploit resources on the continental shelf that are under or fixed to the sea floor. Under international law it is illegal to set up any installation that is a hindrance to the common right of free navigation. One suggested legal definition of the limits of the continental shelf sets it at a depth of 200 metres or 'to where the depth of the superadjacent waters admits of the exploitation of the natural resources'. This last statement would seem to suggest that the ownership of the mineral rights is dependent on the ability to exploit them. In the high seas (areas beyond territorial waters and the continental shelves) the problem is confused by the historic right of freedom of the seas, which includes the use and exploitation of the ocean bed. 'In the deep ocean... the nation that first develops the technical ability will be the first one to claim ownership.'[20]

16
Oil Pollution and the Future Industrialisation of the Sea

Oil pollution has been a problem for the best part of a hundred years but it took the *Torrey Canyon* and Santa Barbara disasters to alert the general public and governments to the dangers. Nobody knows for certain just how much oil enters the sea each year. In 1970 a study group[1] estimated that 0.1 per cent of the world's total crude oil production is lost into the sea every year, since over two billion tons of oil are produced annually the amount polluting the sea could be two million tons. There are two kinds of oil pollution; accidental pollution, resulting from human error, from collisions, strandings or overfilling of tankers, and operational pollution, which is oil deliberately put into the sea as part of a regular procedure.

Accidental pollution hits the headlines; yet most of the oil on our beaches is a result of a deliberate act of pollution.

The deliberate pollution occurs as a result of cleaning out a ship's tanks. The tanks are washed out with water jets and the oil residues are consolidated in one tank. Eighty per cent of the world's tankers retain these residues and discharge them when they reach port. The remaining twenty per cent dump them into the sea. According to international law they are legally entitled to do this in specified dumping zones a hundred miles offshore, but much of it reaches the shore. The laws are shortly to be amended to prevent this pollution and tankers will have to retain their residues until they reach port. There they can be inspected to see that the regulations have been complied with. Tankers will still be allowed to discharge oil containing water effluents, and a strict limit will be kept on their oil content. Tests done by the British government have shown that such effluents leave a light slick which disappears by dispersion within 2-4 hours.

One of the difficulties is that oil is not a single compound but a complex mixture of hundreds of different chemicals with varying properties and toxicities. Its composition varies from one oilfield to another. This makes it extremely difficult to understand its environmental effects.

When oil comes into contact with water it rapidly spreads into a thin layer and the lighter fractions evaporate. In protected areas it may be absorbed on to particles and sink, in the open it tends to stay on the surface. Some oil dissolves in seawater, some is oxidised, and some is broken down by microorganisms. Hundreds of species of bacteria and fungi are able to degrade the hydrocarbons in the oil. Some thrive so well on oil that one of the bright ideas for factory production of protein is to grow yeast-like moulds on oil! The micro-organisms are unable to degrade all fractions of the oil equally. Some of the large hydrocarbon molecules with 'branched' chains persist for long periods as tarry lumps.

Whilst the big spills resulting from accidents have received considerable attention, less has been given to the tarry lumps of oil that have been widely seen floating on the ocean surface. In an area off the south of Italy the American oceanographic vessel *Atlantis II* recorded several thousand such lumps on every square kilometre of the sea's surface. They estimated that there were 500 litres of tar per square kilometre.[2] Just how thoroughly oil is biologically degraded is a matter of some controversy and there are conflicting research findings. Some researchers found more or less complete degradation occurred within a few months. Other workers have found that certain components persist in the bottom sediments.

Many of the most toxic components of oil, aromatic hydrocarbons (e.g. benzene, toluene, and xylenes) and low-boiling saturated hydrocarbons, evaporate easily and a large proportion disappear before slicks reach the shore. However, they may do damage to marine wildlife before they evaporate or reach the shore. The fate of the toxic chemical substances that are in oil is still unknown. Globules of oil are taken up by planktonic animals and the toxic compounds may be taken up into a food-chain. The *Atlantis II* expedition reported finding large amounts of tar in the stomachs of three saury. The saury is a fish which is eaten by porpoises and larger predatory fish. The eating of tar by saury could introduce toxic compounds into the oceanic food webs.

Recent advances in biology have rendered old concepts of what constituted toxicity obsolete or at least incomplete. It is becoming increasingly apparent that the behaviour of many animals is very much influenced by the presence or absence of certain chemicals. Small amounts act as behavioural cues in food-finding, escape from predators, homing and reproduction, and so on. Pollutants may interfere with these processes by removing or even mimicking the behaviour-influencing chemicals. For instance, starfish are attracted to their oyster prey by certain chemicals present in minute concentrations of a few parts per billion. These chemicals may resemble certain chemicals (high-boiling saturated hydrocarbons) that are found in oil. Thus, by blocking the taste receptors and by mimicking natural stimuli, such pollutants could 'fool' a species to its death.

Some of the dispersants used to control oil slicks seem to alter the behaviour of certain animal species. For example, mussels are often unable to re-attach themselves to rocks, Sabellaria larvae and barnacle larvae are unable to settle, the climbing response of top shells and winkles is inhibited, razor shells and heart urchins come to the surface of the sand or mud instead of remaining buried, the hermit crab has been seen to leave its protecting shell. All these changes of habit put the creatures at a great disadvantage. Crabs have shed limbs after exposure to dispersants.[3]

The high-boiling fractions of oil are less likely to evaporate and so they tend to remain longer in the water. Some of these compounds are known to be carcinogens. For example, in 1964 a French marine chemist, Lucian Mallet, found 3,4-benzpyrene, a carcinogen, in sediments in the French Mediterranean. It has not proved possible to confirm whether Mallet's benzpyrenes were natural or due to pollution. However, it seems reasonable to ask what happens to such components of oil spills. Do they find their way into food-chains or stay in the bottom sediments? It is noteworthy that benzpyrenes have been detected in sea cucumbers in concentrations slightly greater than those in the bottom sediments where they feed.

The most harrowing effects of oil are on sea-birds. If their plumage becomes oiled they generally die from loss of insulation, loss of buoyancy and from poisoning after swallowing oil in an attempt to clean their feathers.

The birds which are worst affected are diving birds which spend much of their time under water—guillemots, razorbills, puffins and penguins. Nobody can say with certainty just how many birds are lost but Professor R. B. Clark, who heads the Oiled Sea Bird Research Unit at Newcastle University, believes 'it is likely to run into millions per annum'.[4]

After the *Torrey Canyon* disaster 5,711 oiled birds were found, of which only 120 survived long enough to be released. Of these, it is known that 37 died within the first month after release.

The auk, which is hit very hard by oil, is doubly unfortunate because it does not breed every year, and then produces only a single egg and suffers a normal fledgling mortality of up to 50%. Obviously such a species is in an extremely precarious position, and a steady decline in their numbers has been observed.

The success-rate of attempts to clean up oiled birds varies from species to species. It can be as high as 75% with swans and gulls but with others it may be less than 5%. Various substances are used to clean oiled birds; mild detergents, cooking oil, vegetable fats, fuller's earth and sawdust. If the oil has been successfully removed the birds are kept in captivity until their plumage has regained its natural waterproofing; this can be as long as twelve months in those cases where a complete moult is necessary. The longer they remain in captivity the harder it is for them to be reintegrated into the wild population, which is the object of the operation. The commonest causes of death in treated birds are enteritis, aspergillosis, infective arthritis and kidney and liver damage.

The captivity period should be reduced to 2-3 weeks if a badly hit colony is to be helped, says Professor Clark. This might be possible if a method of cleaning the bird whilst maintaining its waterproofness can be developed. Until this happens the worthy efforts of bird-lovers to protect colonies will be in vain. Professor Clark believes that 'it is likely that measures to reduce the very great loss of eggs and young on breeding ledges would do more to offset losses caused by oiling than any rehabilitation programme for the adults'.

In shallow inshore areas oil can enter a whole range of tidal and subtidal ecosystems. It is emulsified by wave action and its toxicity is increased. It sticks to algae and rocks and smothers many of the small creatures that live on the shore. In the more sheltered areas such as mud and saltmarsh the decomposition of the oil leads to anaerobic conditions, killing saltmarsh vegetation. Normally this vegetation quickly regenerates. But in areas near tanker-unloading or loading points there is often chronic pollution and the saltmarsh is eroded and a valuable wildfowl habitat lost.

Sometimes the attempts to clean up oil pollution can themselves lead to serious forms of pollution. Detergents used to disperse oil are often more toxic than the oil itself and even the low-toxicity dispersants may aggravate the oil's physiological effects by emulsifying it and making it an integral part of the environment.

Another way of getting rid of oil is to sink it with an inert substance such as chalk dust. This method poses less of a problem for the ecosystem as a whole but may prove to be a nuisance to trawling fishing. Burning oil, whilst spectacular (particularly when it is done by Royal Naval bombing as at the *Torrey Canyon* wreck), rarely gets rid of all the oil for it leaves a tarry residue which is toxic and hard to remove.

The use of oil-spill dispersants to solve emergencies resulting from oil-spills is now regarded by most experts as outmoded for it concentrates on the elimination of visual proof of pollution and fire hazard, with little or no consideration of the ecological consequences.

The trend is likely to be towards improved methods of removing oil rather than towards dispersing it. But removal methods are not yet universally applicable and authorities may find themselves in the situation of weighing up the possible ecological damage against the hazards of non-treatment. This is unfortunate, since reliable information about the toxic and ecological effects of dispersant chemicals is still sparse and so we may expect to see blunders occurring for some time yet. One thing is certain—if enough determined effort and cash are put into research and development, products will be developed which are effective at lower concentrations, less hazardous to handle, more readily biodegradable and less ecologically harmful.

Oil Pollution by Accident

The main types of accident leading to pollution are tanker wrecks, oil-well seepages and breaks in pipelines. The classic wreck was that of the *Torrey Canyon*. The advent of the supertanker has increased the possibility of catastrophe; what effect would 200,000 tons of oil have if released after a wreck? In theory, supertankers should mean fewer tankers and therefore lessened chances of collisions. However, they seem to be such unwieldy and cumbersome vessels—they take several miles to come to a stop—that it is doubtful whether this theory will hold in practice. Furthermore, the busiest sea-lanes will need much-improved traffic control to take such vessels safely.

Fears of enormous oil pollution disasters by the sinking of supertankers are increased by the fact that three have already blown up and sunk. Their total value was over £20 million. All three sank during December 1969—the 206,700-ton *Marpessa* off West Africa, the 208,560-ton *Mactra* off East Africa and the *King Haakon*, 222,000 tons, off Mozambique. All the explosions occurred whilst the tanks were empty and being cleaned; they were possibly caused by static electric sparks.

As a result of the January 1969 blow-out at Santa Barbara, California, a local colony of western grebe which had formerly numbered over 4,000 declined to only 200. Over 230,000 gallons spread over an area of 800 square miles, fouling small boats worth five million dollars and sea front property worth two thousand dollars per foot of frontage.[5]

The Governor of California, Ronald Reagan, unsuccessfully attempted to ban drilling in State waters, arguing that present technology is not adequate enough to allow offshore drilling without pollution.

Despite increasing international control it is possible that oil pollution may in the future cause even greater damage to the environment. There are two factors which may make this happen; changing methods of transport—the supertanker and super pipeline—and new oilfields in ecologically sensitive areas. The new oilfields are often offshore wells or in the Arctic.

On 20th April 1970, at Tarut Bay on the east coast of Saudi Arabia, a major oil pipeline broke, spilling approximately 140 million litres of oil into the bay. The spill is not likely to have any long-term ecological effects, but what would be the effect under Arctic conditions in Alaska? What will be the effect on the Great Barrier Reef in Australia, near which test drilling started in 1969? The Great Barrier Reef is already suffering from a disastrous plague of the starfish *Acanthaster planci* which is eating the coral. Oil spills might be the last straw and a unique ecosystem could disappear.

The Dangers of the Arctic Oil Industry

The future development of the oil industry in Alaska is causing great concern to conservationists. Some ecosystems are more sensitive to stress than others. The Arctic is such a sensitive system. Many Arctic species have a low reproduction potential and such species find it very hard to recover from any major environmental blow. In a high proportion of marine invertebrates living in temperate and tropical waters the adult does not develop directly from the egg but passes first through an intermediate free-swimming larval phase. In the event of a local catastrophe such planktonic larvae can migrate from unaffected areas and recolonise the polluted regions. However, in Arctic regions the majority of marine invertebrates develop directly from the egg without going through such planktonic larval stages, and consequently will recover less easily from pollution accidents.

Another factor that has to be considered is the much slower growth and development rates of many Arctic species.[6] The potential dangers are made worse by the fact that it has been shown to be far harder than normal to clean up oil slicks under Arctic conditions.

Oil could be removed from the Alaskan oilfields by either pipeline or tanker. The most favoured method is by pipeline. A £400 million pipeline is likely to be built from Prudhoe Bay to Valdez in south-west Alaska. This pipeline could shift 500,000 barrels of oil daily and will make the oil easily available to American west coast markets and Japan, but not to the east coast regions.

The second method which might be used, particularly to supply the east coast, is to carry the oil in supertankers through the treacherous North-West Passage. The voyage of the giant tanker *Manhattan* shows that such an approach is feasible, but a *Torrey Canyon* accident in Arctic waters, especially likely in such a difficult navigational region, could be an ecological disaster. The Canadian Wild Life Federation appealed for a partial moratorium to halt the uncontrolled acceleration in oil exploration. The slow-down should, they propose, continue until 1974 to allow time for the 'development of less damaging techniques of exploration, development and production and for the ecological research necessary to establish tolerable levels of disturbance in different zones within the Arctic'. It would also give time for further advance in spill clean-up techniques and the development of the necessary staff and organisation to deal with the spills. The costs of clean-up measures in the Arctic would be very high because of the cost of transporting men and materials and maintaining activity under the harsh physical conditions.

The basic ecological knowledge about Arctic wild life distribution and population dynamics and the productivity of Arctic seas is meagre and the danger is that this virgin region will be 'developed' before anyone really knows what the ecological baselines are with which to monitor ecological change.

Future Marine Industrial Developments

Undersea mining will become a major industry before the end of the century and pose a significant threat to the health of the sea. Already, offshore oil, gas and sulphur fields have developed enormously over the last decade; the fields off the American coast produced $1.7 billion worth of oil, gas and sulphur in 1967. The sea also currently supplies industry with manganese, bromine, salt, diamonds, iron, coal, and calcium carbonate. Sea plants are harvested as sources of industrially useful colloids and marine oils. In most cases the offshore production is a mere fraction of land production, but substantial efforts are being made to exploit the mineral resources of the sea.

One undersea mineral resource which has attracted great interest is the metallic nodules. These are generally composed of manganese-iron oxides; they are rounded, earthy-black, layer-like accretions, up to ten pounds in weight and two feet in diameter. These nodules lie on the ocean bottom. Their composition varies from one location to another; for example, sodium, calcium, strontium, copper, cadmium, cobalt, nickel and molybdenum have all been found in nodules. Nodules in the Central Pacific are especially rich in nickel and cobalt.[7]

Millions of square miles of ocean floor are covered with these nodules. It has been estimated that 1,700 billion tons of nodules lie on the floor of the Pacific, and that they contain 400 billion tons of manganese, 16.4 billion tons of nickel, 5.8 billion tons of cobalt, and 8.8 billion tons of copper.

An official American Federal Report[8] recommended the formation of a new independent Federal Agency—the National Oceanic Atmospheric Agency (NOAA)—which would report directly to the President and bring together much of the scattered and diverse governmental and non-governmental programmes in marine science. This recommendation has been accepted by President Nixon.

It further proposed two goals for a national effort in marine technology and engineering. First, to develop the technology to allow men to work on the ocean floor for long periods down to 2,000 feet. The report notes that the most 'productive' region of the ocean occurs at about this depth. Secondly, to develop the capability to *explore* depths to 20,000 feet by 1980 and to *exploit* them by the year 2000. This depth takes in about 80% of the ocean floor.

Already the consortia are being formed to exploit the ocean bed. One such group hopes to produce a system for dredging up nodules from the Pacific at a depth of 12-20,000 feet, and has built a pilot plant for processing nodules. There seems to be very little, if any, attention being paid to the tremendous ecological damage these operations could do in the future if they are not handled and planned with the utmost care.

In a review in *Chemical Engineering*, the advantage of ocean mining operations was said to be 'the generally low cost of waste disposal'. It is this attitude that augurs ill for the future safety of the sea. The enormous increase in costs of such operations over traditional terrestrial mining will create great pressure to cut waste-disposal costs. The mining industry has probably the worst amenity-destruction record of any industry. Vast areas of the world's land are permanently scarred by mining; the effects on the ocean could be equally catastrophic.

Undersea Factories

The work of Jacques-Yves Cousteau and others has shown that already man can live for extended periods under the sea. It seems almost certain that at one stage in the growth-curve of man's exploitation of the sea underwater habitation will become necessary. At first, power-production units and chemical processing plants will be put on the ocean bed surrounded by a seemingly infinite supply of raw material—the sea. The demands for power are growing so rapidly that within, say, fifty years all the ideal coastal power-station sites in advanced nations will be built upon. At this point it will become economic to consider placing the nuclear power-stations on the ocean bed. The growing objection that people feel towards having nuclear reactors in their midst will add to the pressure to drive such activities under the sea. If such developments occur, and they are plausible, growing numbers of people will spend their lives on or under the sea in specially designed dwellings. All these changes will add to the already great pollution burden being placed upon the sea. The designers of underwater cities will have to take into account the fact that it is a 'living sea' and the proportion of capital spent on anti-pollution control will be necessarily high.

17
Pollution in the Soviet Union

In October 1917 the Russian Revolution smashed the power of the landowners, businessmen and industrialists. Throughout the world workers and radicals saw the revolution as the dawn of a new age—a new civilisation. However, despite the enormous advances made by the Soviet Union many of the great hopes engendered by the revolution have never materialised. In the field of pollution as well the record is far from perfect. The Soviet Union industrialised, but only by sacrificing man and the environment.

A major worry is *water pollution*:

'At the present time, approximately 60% of the sewage that flows into reservoirs does not meet sanitary requirements. Rough estimates indicate that each year the USSR fish industry loses from water pollution ... a total loss in excess of 350 million roubles.'[1]

An article in the newspaper *Izvestia*[2] discussed some of the pollution problems that are arising along the River Volga. One of these is water-logged timber littering the bottom after its use as a logging causeway by lumbermen. Another is thermal pollution. There was a large fish-kill near Kostroma State Regional Power Station, a result of a decision to dispense with fish protection measures during its construction. The hot water has now been diverted to another river, the Keshka, where it is hoped that the heated section may be used for fish hatcheries. Further along the river, in the Yuryevets region, fish have been killed as a result of agricultural activities. One Comrade Kruglov, a fish expert of the Yuryevets Fish Conservation Service, wrote, 'the Yuryevets dairy has turned the Yamsk ravine into a sewage canal, from which wastes spill into the river.' Kruglov adds that 'urgent measures' had in fact been taken over four years ago, but without result; nothing had changed. Similarly, the Yuryevets District Soviet passed a decision to stop the unloading of fertiliser on river banks, but these measures also failed to be implemented and the river was full of fertiliser. At Kalinin, also on the Volga, a purification plant is being built. How efficient the installation will prove to be is questionable, as the Kalinin Special Construction Trust is said to be pursuing a 'rouble gross goal'—an indication of progress by the amount of money spent without specifying quality. *Izvestia* adds, 'the amount of roubles expended is not what purifies water'.

According to T. Khachaturov, 'Pure fresh water resources are declining. In many industrially developed regions, an acute shortage of water is already felt. But at the same time rivers are being flagrantly polluted. Three-quarters of industry's sewage flows into reservoirs without undergoing any kind of purification; this kills the fish and turns the rivers into cesspools.'[3]

A great deal of concern has been created by the news that Lake Baikal is now being polluted. The danger of polluting Lake Baikal is not only a direct worry for Russian scientists, but also for scientists all over the world. This is because Baikal is a unique ecological habitat. It is not merely the world's deepest lake (1,741 metres) and the largest fresh water lake on the

Eurasian land mass, but it contains more than a thousand animal and plant species found nowhere else in the whole world. Perhaps the most famous are the unique Baikal fresh water seals.

Baikal is currently threatened in two ways: firstly by a logging and woodpulp industry, and secondly by the development of a modern industrial power complex. Many conservationists have expressed fears that effluent from the pulp and paper mills might create havoc. Despite the fact that official regulations are severe, the measures do not in practice appear to be successful. In the region where effluent is discharged from the paper-mill at Baikalsk the number of animals and plants has been reduced by one-third to one-half.[4] Consequently conservationists have argued that the only safe way to protect the lake is to prohibit any effluent from entering. The effluent from the pulp and paper mills could be carried by a forty-mile conduit over the watershed to the River Irkut, which does not flow into the lake. Such a plan would cost $40 million, which seems a small sum to protect such a unique heritage. However, the pulp and paper Ministry are reported to be strongly opposing the plan.

Ideally, the Baikal region should be declared a conservation zone and no more industrialisation allowed. Perhaps a workable plan would be a carefully controlled combination of tourism and conservation. Baikal has to be protected; if any of its unique life-forms are lost, they are lost for ever and cannot be replaced from elsewhere.

The Caspian Sea is the world's biggest inland sea. It is also so polluted that Russian scientists claim that it is heading for catastrophe. Dr A. G. Kasymov has said, 'If pollution of the western part of the middle and southern Caspian Sea continues as it is now, the sea can be expected to be transformed into a dead sea, not only unsuitable for habitation by fish and other food animals, but also for the needs of technology.'[5]

The Caspian receives large amounts of oil and sewage pollution. Most of the oil comes from seepages from underwater oil wells and from tankers. The seepages are very numerous; in some places, for example, at Neftyanye Kamni Island, Kasymov reports that 500 metric tons of oil escape from seepages every day. Much oil enters the sea from broken pipelines and poor loading procedures. Oil production in the area is increasing, and so the problem could get worse. In some places the surface of the sea is covered by a thick layer of oil, and the bottom is impregnated with petroleum products. The reason for the oil pollution, says Kasymov, is that 'the industry is not very enlightened', and he adds, 'by the end of the century, the Caspian Sea will have irreversibly lost its value as a productive source of fish.'

During the last thirty-five years, fish catches have dropped by almost two-thirds to 110,000 metric tons. This is thought to be due to the dropping water-level in the Caspian, which is gradually drying up; the effects of a series of hydro-electric stations along the River Volga, the largest river entering the Caspian; and pollution by local towns and industry.

The Soviet Union is famous amongst other things for its caviar. This is the roe of the sturgeon. The sturgeon, which spends much of its life in the Caspian, is on the decline. Caviar is not only a great delicacy, but a good source of foreign currency. The combination of the decline of the sturgeon and the requirements of the Soviet balance of payments means that the average Soviet citizen rarely sees caviar in the shops. Dr Kasymov reports that the Central Scientific Research Council of the Fishing Industry has announced that annual sturgeon losses now amount to 5,000 metric tons a year, and that 'If pollution on this scale continues, sturgeon and sild will soon be completely wiped out and the value of the Caspian Sea to the fishing industry will be reduced to zero.' Prospects are so bad that research into the production of artificial caviar has started. Maybe in ten years' time we shall be regaled by television commercials showing housewives nibbling wafers thinly spread with caviar substitute and breathlessly announcing that they cannot distinguish it from the real thing.

Air Pollution

'The state of the air in many cities is a source of concern. Of the total number of industrial enterprises that are acknowledged to be a source of air pollution, only 14% are fully equipped and only 26% are partially equipped with purification facilities.'[6]

Pollution of the land

'The situation with respect to the use of farmland is no

better. According to estimates by the Dokuchaev Soil Institute, losses due to soil erosion constitute 3.5 billion roubles per year ...'[7]

Like their western colleagues, the Soviet agricultural workers have concentrated on narrowly based cost-benefit studies of pesticide use, based on the cost of pesticides compared with the value of increased yields obtained from their use. Professor N. Gladkov and Mrs L. Voronova of the Laboratory for the Conservation of Nature, and others, have pointed out the necessity for taking account of the cost of environmental pollution. Others have argued that this could only be done if such cost-benefit analyses were the concern of the whole Gosplan (the State planning commission) rather than of the agricultural ministry alone.[8]

Mrs L. Voronova, who is the Soviet representative on the Toxic Chemical Committee of the International Union for the Conservation of Nature, has stated,

> 'when they investigate a new pesticide preparation they only look to see if it will harm domestic animals and birds. They don't examine the effect on the whole environment. Only a ridiculously small number of people study the "chemistry-nature" problem here. You can count them on your fingers. In my laboratory in Moscow, five men, in the Ukraine—one, in the Tadzhik ASSR—one.'

> 'In the Stavropol, Taurbov, Ruazan and Ulyanov districts, a massive poisoning of wild and domestic animals has come to light. In Tartary, the Karelian ASSR and the Mogilev, Moscow, Odessa and Ivanov districts, poisoned hares, elks, quails and partridges have been discovered. Not every instance is brought to light or investigated. The scale of the disaster grows every year in proportion to the volume of chemical treatment of the land.'[9]

Squandering of Natural Resources

'To this very day our nation still has not created the conditions that would make it economically disadvantageous for enterprise collectives to make irrational, uneconomical use of natural resources.'[10]

'Unfortunately certain of these measures (legislative and government) do not prove to be sufficiently effective. Enterprises in the extractive industry bear no liability for minerals not extracted from the ground and no-one is responsible for either the valuable new materials that are thrown on to rubbish heaps or for unutilised wood scrap. Sanctions for polluting water reservoirs and the atmosphere are imposed on enterprises and are in fact paid for from government funds, while those guilty of destroying natural resources are punished only in rare cases. There are too few inspectors in the sanitation-engineering and natural resources protection service. They are not invested with the necessary powers... Clearly stronger administrative measures must be taken and liability for violations made greater. Fines should be extracted not only from the profit of the enterprises, but also from the wages of enterprise heads, with whose knowledge unpurified waters are discharged and the atmosphere is polluted.'[11]

'This problem (of natural resources) has become particularly acute in connection with the implementation of economic reforms. If profit becomes the chief indicator of the effectiveness of an enterprise's work, enterprise collectives are more "interested" than previously in the irrational exploitation of natural resources. Hence the contradictions between the social necessity of making rational use of natural resources and the lack of interest of enterprise collectives in doing so, are exacerbated.'[12]

Industrialist Attitudes

Like their capitalist counterparts, some of the industrial managers do not like to admit pollution exists.

V. N. Platonov, acting president of Machinoimport, told delegates of the Delaware Chemical Industrial Trade Mission, 'We have a general plan for controlling air pollution from oil- and coke-burning plants through the installation of electrostatic control equipment. In the Soviet Union, plants are not put into production without the proper pollution control equipment.'[13]

The existence of a pollution problem in the Soviet Union and in other communist states is used as an argument for the theory that pollution is simply a problem of industrialisation itself and not, as I have argued, a result of the capitalist mode of production.

There are two points that need to be stressed in answering this argument. First, I have attempted to analyse the reasons for a pollution problem within capitalist economic systems. Secondly, I have not suggested that the solution to the environmental problems is to establish a replica of the Russian economic system. However, it is necessary to explain why it is that large-scale pollution occurs in the Soviet Union and similar states.

The Soviet Union was the first country in which the means of production were nationalised and operated according to a plan. This was the major factor behind the fantastic pace of industrialisation and education that raised Russia from a backward, largely traditional, peasant society to become the second most powerful state in the world. If the economy was planned for forty years, why was it unable to plan adequately for pollution control? One needs to point out first of all that the industrialisation of a backward economy, forced to depend on its own resources, isolated, surrounded by enemies, devastated by civil war and then by World War II, could only be done by tremendous sacrifice and hardship. That there would be under such conditions a strong temptation to put growth before pollution control is understandable. But that is not the only cause of, or even the fundamental reason for, pollution in the Soviet Union. The basic reason for their pollution is the distortion of planning by bureaucracy.

The development of the bureaucracy prevented rational planning. Plans were conceived by the upper echelons of a bureaucratic hierarchy. The implementation of the plan was by factory managers responsible not to the workers within the enterprise, or in society at large, but to superiors in this bureaucratic hierarchy. The working class and peasants were not allowed to play a constructive role in the drawing up and implementation of the plan. Rule by a minority, instead of by mass participation, will never produce the quality of information necessary for the total planning of the economy.

Because there were enormous shortages and because there was rigid centralisation instead of mass participation, the bureaucracy obtained for itself many privileges. The economic plans began to be distorted by the material and political needs of the bureaucracy rather than corresponding to those of the nation as a whole. There was an emphasis on material incentives which tended to exacerbate anti-social features of plans. A Soviet manager's 'success' was assessed on his ability to meet the letter of the plan, and he was given monetary incentives to do just that, thereby creating a situation in which pollution problems were inevitable. The accounting systems of such bureaucratic planning were very mechanical, based on production and growth indices with little or no attention given to the third-party costs of production.

In capitalist society private property relations prevent the development of social rationality, making comprehensive planning impossible, whereas in the Soviet Union the development of social rationality is retarded by bureaucratic planning. To get rid of such bureaucratic hindrances to satisfactory social progress the Soviet people must have the right to criticise and to hold free elections.

PART THREE

The Techniques of Pollution Control

18
Pollution Control Technology

In this chapter we shall take a brief look at some of the available techniques of pollution control.[1]

Take a hypothetical, but by no means unusual, polluting chemical plant. Nearby is an estate of several hundred people. We will give this plant the name of RentaSmell. It is a huge, ominous structure, surrounded by a wasteland of blackened earth, whose monotony is relieved by dumps of strange material, sulphuric yellow or vivid white, near which children are not allowed to play. Between the works and the river is an estate of tiny houses, submerged in a cloud of dust, fumes and smoke. The houses are blackened and crumbling by the action of chemicals on the brickwork. The people are pale, grimy, anxious and unhealthy. Odours from the river are stronger on sunny days, and the dust is drier in the throat, so in some ways the grey days are best. The people don't go down by the river, and children don't play along the banks. Mounds of foam float gently on the black waters, and the scum which laps at the mud-banks leaves multicoloured rims of creamy deposits and slime.

Let us first turn our attention to the river that RentaSmell has killed. No-one can fail to notice floating, suspended substances of curious colour, the scum along its rim, and the foul odours that strike us from a hundred yards away. We know this is the work of RentaSmell, because before the river reaches the works, it is fairly normal and fish live in it. The effluent gushing from spouts in the wall overhanging the river is composed of a variety of solids, liquids and dissolved chemicals. We will assume for the sake of our story that these are known and identified. Furthermore, we will assume that they must be treated on the spot, as the local sewage-works is not prepared to take on the job.

The effluent could be treated in the following manner: Firstly, suspended solids could be removed from the liquid waste. The methods used for this primary treatment include screening, sedimentation and flotation. Screening removes large floating particles which might damage equipment, whilst sedimentation is the settling of these particles by gravity—this is done in tanks or lagoons. Flotation, which gives the highest separation rate, is the opposite of the sedimentation process. By attaching fine bubbles to the solids, the combination rises to the surface and can be removed.

How we treat the remaining effluent depends on whether it can be broken down by bacteria or not. In the case of RentaSmell, their effluent contains many substances that cannot be broken down in this way—for instance sulphuric and nitric acids, waste alkali, and toxic chemicals—all of which have to be removed before any biological treatment can be given. An important part of our chemical treatment would be the neutralisation of acid and alkali waste. Our plant has both acid and alkali waste so we will be able to blend them to get a neutral stream. Other chemicals might be treated by oxidation and reduction processes.

Some of the remaining effluent could be treated biologically, using nature's own method, the action of micro-organisms. Given the appropriate conditions, micro-organisms can convert organic waste to gases and water. In our case, seeing that we are treating industrial waste, care has to be taken to see that water temperature, pollutant concentration, and possible nutrient deficiency are considered and necessary corrections made. In particular, we should have removed any toxic substance that might kill the micro-organisms.

Maybe some of RentaSmell's problems are so complicated that some of the latest techniques now being developed will have to be employed. Suppose, for example, that our effluent contained some very stubborn organic material that was highly persistent and so could not be broken down by the biological treatment. This might be dealt with by the absorption method. This consists of passing our effluent through a bed of activated carbon granules which removes most of the polluting organic substance.

Another promising process, which may solve RentaSmell's more refractory problems, is reverse osmosis. To understand this method we must first of all say a word about osmosis. Imagine a situation where two fluids, with different concentrations of salts dissolved in them, are separated by a membrane through which the water molecules can pass but not the dissolved substances. In such a situation, we find that molecules of water pass through the membrane from the less concentrated to the more concentrated solution. This will go on until the solutions on both sides are of the same strength. This process is called osmosis. The direction of the osmotic flow can be reversed, when pressure is exerted on the side containing the more concentrated solution. This reverse osmosis will cause the water molecules to come out of the compartment that contains the most concentrated solution. As a pollution control device, this system reverses the traditional approach since it takes the pure water out of the waste rather than taking the pollutants out of the water. In theory, this allows a near-perfect cleaning operation. It seems that the perfection of this process depends very much on a significant improvement of the membrane design.

Most of the treatments discussed above end up by concentrating the pollutants into a sludge. The final aim is then to remove the liquid and stabilise the solids. The major operations involved here are vacuum filtration, centrifugation and concentration. Even after the completion of these, there is still solid waste remaining.

Perhaps the cheapest way of disposing of this is the multiple hearth incinerator. The ash remaining is sterile and inert and can be used for filling land or even for the manufacture of building bricks and concrete blocks.

Our next problem is air pollution. How can we lift the haze in which these people live and purify the air by removing the irritants which give young children chronic coughs? The principal methods include dry mechanical collectors, gas scrubbers, electrostatic precipitators and the use of tall chimneys. We will not go into the details of these here as we are more concerned with general principles than compiling a textbook of pollution control methods.[1]

If for the sake of argument we say that our plant is emitting sulphur dioxide from flues, what can be done to prevent this pollution? According to a report commissioned by the National Air Pollution Control Administration in the United States,[2] there are sixty-five possible methods of controlling sulphur dioxide emission, and most of the promising methods involve extended conversion of the stack gas and/or chemical absorption and neutralisation. The method that the report cites as receiving the most attention is the dual-absorption method.

In this method, sulphur dioxide is converted into sulphur trioxide which can be converted into sulphuric acid. This method can eliminate 99.5% of the sulphur dioxide in smoke. If, in the case of RentaSmell, the sulphur dioxide comes from the burning of fuel, then we can improve the situation by using fuels of low sulphur content.

A constant background to the other discomforts besetting the people who live near RentaSmell is the throb and drone of machinery, and high-pitched whistling noises.

In order to reduce the noise from our plant we must find out the actual sources of noise and the path of travel. In general there are two main approaches to reducing the level of noise. Firstly, one can prevent or reduce noise at its source. For example, one of the most unpleasant sounds is the release of high-pressure air through jets or nozzles, as this is usually high-pitched. By improving the nozzle design, muffling the air exhaust opening, and altering the angle at which the air hits a surface a noticeable reduction can be obtained. Secondly, one can put an obstruction between the noise source and the people affected by it. Sound-absorbent barriers can often help to reduce the effect of a particularly noisy machine to an acceptable level for the workers and local inhabitants.

Another characteristic of RentaSmell and its neighbourhood is a particularly offensive smell, so pungent that its penetrating sickliness can be perceived, not only above the general smells of soot and dirt, but even through the haze of long habitation. The culprit in this case is ethyl mercaptan, but it could easily be any one of a combination of various malodorous, untreated by-products. The state of the science in this field is such that each problem has to be approached individually. The main approaches are: burning the vapours to give non-odorous substances, condensation of the gaseous substance into a liquid, absorption, or pushing it out of a very tall chimney. The last does not really deserve the name of pollution control, but in the case of pungent odours which are otherwise reasonably harmless, this may be acceptable.

RentaSmell, in common with many factories, accumulates piles of solid waste. This waste could be recycled and thus converted into useful material. For example, cellulose waste can be chemically converted into useful products by pyrolysis. The possibilities of re-cycling will be examined in the next section. RentaSmell need not despair even if there should appear to be no possible use for this solid waste. Waste can be compacted under high pressure to give blocks of high density. These are then covered with chicken wire, or thin sheet metal, and dipped into asphalt, concrete or plastic. They can then be used in building houses, dykes, walls *et cetera*. But unless such projects are carried out with real care, they can lead to worse trouble. In America, on at least two occasions, sand contaminated with radioactive material has been used to build houses.

Thus most of RentaSmell's pollution problems, which are making the lives of hundreds of people a misery, reducing the health and stamina of children who live nearby and demoralising and poisoning the adults, are technically solvable. An intensive cleaning-up programme could clear the wastes and dumps, so that grass and hardy plants might begin to grow. Dirt and acid fumes would be cleared from the air and the coughing of children would no longer be heard at night. The mothers could begin to get some satisfaction from their homes; it would be worth having them painted now, as the paint wouldn't immediately peel off, as it did last time. It would be worth putting up pretty curtains, as they would no longer rot away before the end of a year. The place would not become a paradise overnight, as there would be many other problems of a more social nature to be cleared up, but life would be more worth living. What prevents places like RentaSmell from undertaking some measures to remedy the damage inflicted on their neighbours? The technical means are there, why aren't they used more often? In later chapters we will analyse some of the reasons, stemming from our current social and economic structure why the RentaSmells of the world are not going to disappear overnight.

Recycling of Waste[3]

The traditional methods of getting rid of waste and trash are burning, burying and carting. These form the basis of the main methods in use today. In most industrial countries, up to 90% of trash is disposed of by open dumping or tipping, burning and land-filling. A land-fill is, simply, tipping the rubbish into a hole or into marshland until it is filled up. By this means a lot of useful land has been developed; rubbish can, for example, be piled into mountains and turned into ski-slopes. However, most of the best sites have now been filled up. In the north-west of England it is said that the quickest way to become a rich man is to buy an old mine-shaft and charge people for permission to tip rubbish down it.

As a result, more efficient methods are being looked for. One way of slowing down the rate at which you fill up your dump is to incinerate, which greatly reduces the volume of the trash. The only problem with this is that you may cause air-pollution.

The only real way to solve the problem is re-cycling, that is, the reclamation of the materials contained in the trash so that they may be used once again. The beauty of this approach is that besides getting rid of the trash, waste-disposal may become self-financing or even profitable. Finally, and most important, it would slow down the rate of exploitation of existing virgin resources.

In the ideal situation, rubbish would be separated back into its component parts which would be shipped back to the original manufacturer for re-use. In actual practice, the authorities have many reasons for not being prepared to do this, of which the chief one is the cost. Most of the money at the moment goes towards *collecting* the rubbish and because local authorities are always under pressure to keep down costs they dispose of it as cheaply as possible. On the basis of the narrow economic criteria used in our society, land-filling is the cheapest way to get rid of waste, until you begin to run out of suitable sites. After that, authorities may look at incinerating and composting as possible methods. In a recycling system, trash could have three uses. It would be a mine of useful material such as iron, tin, paper; its organic component could be turned, via composting, into food. And those materials which could not be separated could be turned into fuel, either by simple incineration, or by chemical conversion into oils. Unfortunately the economic and social barriers are far higher than the technical barriers. Let us have a look at some processes of rubbish disposal already technically available.

Incinerators. A modern incinerator can be more than just a place where rubbish is burnt. More sophisticated modern models incorporate anti-pollution devices, and the heat that they give out may be used to drive turbines, or to provide heating or hot water. After this burning process one is left with an incombustible residue composed chiefly of metal and glass. Economic means of separating these residual components are the subject of current research. The US Bureau of Mines have already developed a comparatively simple process using screens, magnets and other devices to separate the components. It is claimed the process might produce enough salvageable material to break even. The economic value of the system would be enhanced if it could be combined with a power-generating incinerator.

Pyrolysis. This process is a development of the old charcoal-burning in which wood was heated in the absence of air, producing, besides high-grade carbon, a wide range of gaseous organic chemicals that could be distilled off. When this system is operated with trash it provides, in addition to the residual ashes, substances such as methyl alcohol, acetic acid and acetone,

The ash can still be mined for its metal and glass content. In past times, pyrolysis of wood was the chief method of obtaining these chemicals. Unfortunately most of these substances are said not to have a large enough market to absorb all that could potentially be produced by trash pyrolysis. This means that given the present structure of the chemicals industry, many of these chemicals wouldn't find a market, except possibly as fuel. But in a socially rational economy it might not prove impossible to integrate industrial activity with rubbish disposal.

There is another process related to pyrolosis in which the trash is reacted with carbon monoxide at high temperature and pressure to produce an oil similar to petroleum. The Firestone Rubber Company has developed a pyrolytic process which can cope even with the hitherto intractable problem of old car-tyre disposal. The process gives 45% carbon residue, and the remaining 55% is a mixture of liquids and gases resembling petroleum. Pyrolysis can convert waste plastics in a continuous process into hard and soft waxes, greases, adhesives and tars, says Dr I. E. Potts of Union Carbide. The hard waxes could be utilised in such products as polishes, printing inks and polymer lubricants. Other pyrolised plastic might be used in asphalt, as humus, or as a source of carbon in micro-organism synthesis of proteins.

Composting is an industrialised form of the gardener's compost heap. The trash is first of all screened to remove metal, glass, plastics *et cetera*, and then the remaining organic component is shredded and placed on heaps or rotating drums; over a period of weeks the material is allowed to decompose. It can then be added to the land as manure, to improve soil structure. It is not a fertiliser in its own right, but has potentialities for maintaining the health of the soil. Many farmers in the English Midlands have recently been made abruptly aware of the importance of soil structure as they have experienced the first signs of soil destructurisation. In some cases whole areas of grass have died. In one notorious case, the soil was so compacted that roots just could not penetrate it. This was caused by the excessive use of heavy machinery and failure to observe adequate crop rotation.

Most American attempts to set up compost plants have proved a failure, however, because of the lack of a market. The usual problem was that they were established too far from agricultural areas, and therefore there were high transport costs which pushed the price above that of artificial fertilisers.

Salvage is the old-fashioned name for re-cycling. It is still exercised with materials which are worth enough to make it a commercial proposition. Twenty per cent of American paper production comes from waste-paper. However, most of this waste does not come from household trash. It comes mainly from commercial sources, and is collected from office and factory packaging departments. There is little doubt that the re-use of paper could be stepped up if better use were made of municipal trash. The paper in trash is hard to separate, but it can be done.

One of the main reasons that paper manufacturers give for not using more salvaged paper is that it is contaminated with all kinds of additives such as wax, plastic, and pigments. New methods of pulping trash so as to allow cellulose fibres from paper to be skimmed off and collected for use are being developed. These methods may open the way to producing a purified form of fibre that can compete with virgin wood pulp. This would be an important breakthrough, as paper is potentially the most valuable component of municipal trash. A lorry-load of municipal trash can contain as much cellulose fibre as a tree newly felled by the lumber company. It is technically possible to save the rapidly dwindling forests of the earth, and as far as environmental quality is concerned there would be yet another bonus—for we would reduce the activity of the wood-pulp industry which pollutes so much of the earth's wildernesses and water supply.

Another possible use for cellulose materials is to convert them into protein. Louisiana State University has a pilot plant which is turning cellulose waste into a low-cost, high-protein food. This developed from the discovery of a micro-organism that breaks down cellulose into a nutritious protein.

Glass and metals remain after incineration. They can now be easily separated out, iron by magnetic devices and glass by powerful blasts of air. Coloured glass can even be separated from non-coloured glass, using high-intensity magnetic fields which can sort the colours according to their iron content. The problem remains of what to do with these materials once they have been separated. The iron and steel industry use a great deal of scrap, although tin cans do pose problems—for they have to be de-tinned. There is, however, no technological barrier to re-cycling tins; the problem arises because de-tinning costs money which the steel industry finds it uneconomical to pay. This could be circumvented by coating cans with substances other than tin; special resins are already available for this use.

Glass is not very valuable—it is worth about fifteen dollars per ton—so various people have considered ways of disposing of it. One group tried grinding it down and using it as a replacement for sand in building materials. In Toledo, Ohio, an experimental stretch of road was paved with a mixture of glass and asphalt.

It is quite possible that certain things may prove incapable of being recycled no matter how hard we try. An engineer and a geologist recently put forward the interesting idea that these wastes could perhaps be disposed of within the planetary interior. There are certain zones where the surface material of the earth is drawn into the mantle, and from there disappears via 'tectonic sinks' into the interior. These sinks are 'apparently so capacious as to take care of any conceivable quantity of material ... its effect on the environment would not be cumulative. The ultimate effect to be expected would be the discovery of a distinctive, and perhaps puzzling, suite of metamorphic rock.' In this fashion we would have achieved the ultimate biogeochemical cyclation.[4]

Collection

One of the perennial arguments is whether it would be best to collect the various components of trash separately or to rely on developing a centralised system. Commercial scrap-dealing, which is worth $8 billion a year in the USA, bases itself on a specialised form of collecting.[5] It is responsible for most of the re-cycling that does occur—recycled material is used in 30% of aluminium production, 45% of copper and brass and 52% of lead. There was a time when the salvage industry was a whole series of specialised trades.

Henry Mayhew, the nineteenth-century social commentator,

describes how the filth of London was recycled by a complex detritus sub-culture of street-buyers, finders and collectors. Most of these scavengers of Victorian London, with wretched lives and picturesque names, such as 'mudlarks', 'bone-grubbers', 'cigar-end collectors', 'dredger-men', have now disappeared.[6]

Re-cycling of municipal household rubbish is especially low. Centres should be built to deal with it. The fact that private industry has not found it profitable should not be a deterrent, as public services are presumably significant more for social reasons than for making money. Most householders would be quite prepared to keep paper or other things separate, if means were provided and they knew it was for a good reason. It would not be difficult to design receptacles to take different materials such as cans, waste food, paper, and glass—possibly with different coloured plastic bags.

Many governments are making the right noises about recycling. The Resource Recovery Act, passed by the United States Congress in 1969, lays special emphasis on re-cycling technology. Many politicians bandy about the concept of re-cycling. A typically optimistic version has been provided by Hollis M. Dole, Assistant Secretary of Mineral Resources in the US Department of the Interior. He describes solid waste as 'our only growing resource... The resource potential of such "solid wastes" is so obvious that you are forced to wonder why we call them "wastes". There they lie... needing only a few technological refinements or perhaps a slight restructuring of the scrap industry to bring them back into the manufacturing cycle'.[7] But does this really bear any relationship to the present political and economic reality in the United States? A survey made by the Institute of Solid Wastes of the American Public Works Association recently found a trend *away* from salvage of recoverable items from municipal refuse.

In a review of solid wastes schemes the journal *Chemical Engineering* put its finger on the heart of the contradiction. 'Many technically feasible answers are economically disastrous.'[8] It is obvious that more than technology alone is required.

The waste-material industry must be treated as a basic industry such as mining, timber or agriculture. This would require special subsidies and the expenditure of a great deal of money. Even this would not be enough, for the disposal industries and the manufacturing industries would have to be re-designed as an integrated unity. Before new industries were allowed to be introduced, techniques for their re-cycling would have to be ready for action at the same time as production.

The Pollution Control Business

Let us just look at the amounts of money that are going to be spent in the field. In 1969, the journal *Environmental Science and Technology* estimated that the 1970 expenditure on all forms of pollution control in the United States would be 1.3 billion dollars.[9] About 250 million dollars would be spent on control equipment, about 50 million on control instruments and nearly 600 million on the necessary chemicals. On top of that is the talk of spending tens of billions of dollars over the next five years or so. In Europe also, the pickings look good, e.g. in the field of plant and equipment for water pollution control. That market will expand from 750 million dollars to an estimated 1,250 million dollars by 1976.

Obviously with the large and diverse giant chemical companies pollution control will be a growing but not commanding part of their business. But there will doubtless be a growing number of smaller firms which will specialise in the pollution control business. Research-Cotterell is an example of such a firm. It claims to be the world's largest company devoted to pollution control. Its revenues have soared in recent years, from $11.1 million in 1965 to over $58 million in 1969.[10] The pollution control business ranges from the big chemical corporations like Du Pont and ICI through specialists like Research-Cotterell down to small, one-man consultancies.

The pollution control consultants are another group who will enter a boom period as a result of these changes. They are very often one-man, or two- or three-man, operations who tender their expertise to companies faced with meeting some new anti-pollution regulation. Generally they deal with the small companies who cannot afford their own specialist staff. There were at least 160 such American consultants in 1969, and they will undoubtedly grow to a much larger number. Many of them have, in keeping with the current fad, picturesque names such as 'EnviroTech Inc.', 'Oceanonics Inc.', 'Resources Research Inc.', and so on. Usually such consultancies do not manufacture or even sell equipment, but examine a company's problem and advise on the best systems to deal with it.

With the sweet smell of profit in the air, some of the world's stock-markets are beginning to take an interest in pollution. In early 1970 *Chemical and Engineering News* ran a most interesting article entitled 'Wall Street Focusses Attention on Pollution Control Stocks in an Otherwise Gloomy Year'.[11] The article described a speculation boom in companies specialising in pollution control which occurred at the end of 1969 and early 1970, when the general stock market position was poor. The article said, 'stocks of companies in the anti-pollution business have been Wall Street's darlings. Since the play began last fall, prices of many such issues have climbed by 50% or more... Clearly, a lot of people, institutional investors as well as individuals, expect a lot of companies to make a lot of money in pollution control.' Stockbrokers were quoted as saying, 'although you have to pay the price if you want to play the game, prices are not dangerously high on the whole, because these issues are likely to keep their glamour image for a long time... Prices of some of the stocks could go to a hundred times earnings if we get a strong market.'

A high price-earnings ratio indicates buoyant market confidence, for it means that investors are prepared to pay a high price now in anticipation of higher earnings later. Here are some of the price-earnings ratios for a few pollution control companies on the Wall Street Exchange in early 1970: Research-Cottrell 85, Ecological Science 50, Betz Laboratories 70, Aqua-Chem 46, Zern Industries 46. Obviously there was an awful amount of speculative optimism around.

Not everyone on the stock exchange is so optimistic, and they were quick to point out the highly speculative nature of the interest in these undertakings, and thought many of these stocks were over-priced. They pointed out that despite all the talk of vast sums of money to be spent on pollution control, only part of it would go to the equipment makers and the rest would go to builders and construction contractors. Furthermore, it was pointed out that a lot of the equipment being used is of a rather mundane nature based on 'mature technology', so the field is highly competitive with low profit margins.

One of the ironical things about speculation on anti-pollution control stocks is that it cuts both ways. Firms which had a vested interest in certain types of pollution dropped in price on the stock exchange. For example, Ethyl Corporation, which is being hard hit by the anti-leaded petrol regulations, had in early 1970 dropped to its lowest stock price for seven years.

In 1970, at the height of the bull market for pollution business shares, all government reports appearing on pollution boosted such speculation; however, this is no longer the case because more of them are making the distinction between pollution control and waste disposal. The former renders harmful materials harmless, and—ideally—re-cycles them for useful purposes, whereas the latter may often mean that obnoxious materials are merely moved from areas where they are banned by legislation to places as yet unprotected by legislation. As it happens, waste-disposal is one of the most profitable parts of the pollution business. Thus any criticism of it causes the stock exchange to act more warily.

In February 1971, a report by the British Government Royal Commission on Pollution chose as 'priorities for action' the disposal of toxic wastes on the land and in the sea, and looked for stiffer legislative control against indiscriminate dumping. This report[12] has been seen by some as a major threat to the waste-disposal business, which will, for a price, get rid of obnoxious wastes for industry. They often do this by tipping it on to land sites or into the sea where they may well create new pollution problems. Some financial experts have predicted that the report will shake stock exchange confidence.

The *Sunday Times Business News* was quick to seize on this aspect of the Commissioners' report, and to speculate what effect it might have on Purle, the leading British specialists in industrial waste disposal. Purle, said the *Sunday Times*, was 'at one time one of the most spectacular performers on the stock exchange'. 'Cesspits, rubbish dumps and getting rid of intractable chemicals at sea are what the unglamorous Purle business is all about. So it is ironic that the Commissioners ... now pose a threat to Purle's best money-spinner.' This was a warning to all speculators in the field, since, as the *Sunday Times* pointed out, 'the

Purle share price is supported by little but growth record and profit expectations'. They predicted that until the Purle Board was able to reassure its shareholders that they would be able to cope with possible changes in the legislation governing the disposal of waste, 'the shares will hang fire at the best...'[13]

The cynical speculative spirit surrounding this industry was summed up in a recently published statement: 'The pollution control industry is going to cost you—directly or indirectly—a lot of money during your lifetime; you might as well try to share in the growth.'[14]

19
Legislative Control of Pollution

The law is said to defend and guarantee the rights of the people. But what are the rights of the people?

Many people believe that our society can protect persons, and ought to put people before things. They believe that mankind possesses certain inalienable rights. To list some of those rights which are concerned with the environment, we have the right to:

a clean and beautiful environment
the consumption of uncontaminated water and food
breathe clean air
live in clean uncrowded housing.

In short, to live in and enjoy wholesome, healthy surroundings. It is obvious that the law is not *primarily* concerned with such rights. If it were, then with the first signs of pollution, or of potential pollution, legislation would be quickly enacted. However, in a capitalist society the most fundamental and basic right is the right to private property, which in turn leads to the right to extract surplus value from the labour of others, the right to plunder the irreplaceable natural resources of the planet. The insistence upon this right as the fundamental basic right, the foundation of our society, leads to pollution.

In most cases legislative activity can be divided into two phases. That enacted in the first phase is scattered and piecemeal. Because the health of the citizens and the environment is not the law's primary concern pollution is seen in terms of isolated problems. Each is dealt with by a series of separate enactments whose texts are spread over a wide variety of legislative provisions. These may be found in local legislation, public health laws, and legislation relating to premises used for dangerous, unhealthy and noxious trades. Certain specific causes of pollution have become the subject of special provisions.

The growing scale of pollution and the increasing variety of damage it can inflict on industry, as well as on man and nature, have led to tardy efforts to provide an overall solution. There is the beginning of an effort to replace the piecemeal codes of the past by more generalised legislation. Legislation has resulted not only from the obvious need for it but also because of pressures from sections of the population enjoying a higher standard of living, who are less prepared than used to be the case to put up with the inconveniences of pollution.

In most countries there exists a general provision to the effect that no one, whether at his place of work or on his property, is entitled to endanger his neighbour's health or cause an excessive nuisance. Such provisions usually exist not only in private law but in many cases also form part of the general health and factory legislation.

In the long run, however, it has been found that such provisions are not sufficient. Their main shortcoming is that they can only be effective when the victims complain to the courts or to the administrative authority. Many serious cases go unremedied because affected parties do not wish to incur the risks and expenses of litigation. The victim may even be economically dependent on the source of emission, as, for example, where his home is situated in the vicinity of the factory where he works.

Again, all the legal requirements necessary to uphold a complaint may not be fulfilled, or the complainant may not be able to trace the offending source. Nor are complainants usually equipped to assess adequately the danger of pollution. Only a specialist is in a position to make precise pollution measurements and express a valid opinion, and even then only after careful study. Therefore authorities should not wait for complaints, but take preventive action in order to limit the emission of pollutants.

In our society the whole problem of pollution control is bedevilled by the conflict between the necessity for profit and the need for a healthy population. Absolute environmental purity is regarded not merely as technically unobtainable but as economically unnecessary. Means are devised for ensuring that the concentration of pollutants does not rise above a specified level, but the health experts are by no means agreed on what the danger thresholds are for the various pollutants. Here we must mention the concepts of *hygienic* and *sanitary* standards. Hygienic standards ensure the biologically optimum environment for man, whereas sanitary standards are a *compromise* between biological, technical, economic and political requirements. No country yet operates any hygienic standards for pollution.

Solutions vary a great deal from one country to another. There has been a tendency in some countries to fix the tolerance limits for the different toxic substances at the 'lowest possible' level. Whilst this sounds progressive, in actual fact it operates against the general public in favour of the polluter. It is a concession that adapts to the *current* state of the industry. A truly progressive approach would set as a goal standards which would bring forth the spirit of inventiveness by going beyond the current state of the act.

Pollution control legislation appears a very liberal act, the act of a socially conscious legislature. However, because social legislation within our society reflects the self-interests of a social minority we need not be surprised that the anti-pollution legislation is generally a response to crisis and disaster rather than genuine social concern. It tends to be remedial rather than preventive, that is, measures are taken after pollution becomes apparent rather than before.

In the following sections we shall sketch the existing legislative situation in certain countries and indicate possible developments within the system.

English Legislative Measures[1]

Common Law, as distinct from statute, provides little *practical* assistance in the fight against environmental pollution. In theory it could because the only criminal offence committed by polluting the environment is that of public nuisance, which includes any act which 'endangers the life, health, property or comfort of the public in the exercise or enjoyment of rights common to all Her Majesty's subjects'. This could cover many forms of pollution; in practice it is very little used.

Under common law, civil rather than criminal actions have been more common. However such actions, taken together, are ineffective in the battle against pollution because they only protect the interests of a successful plaintiff, whose interests have been affected to such an extent and whose income is substantial enough that he is willing to incur the expenses and risks of litigation. Pollution is a problem of such urgency and universality that measures to combat it cannot be left to the resources of the private individual.

Therefore, in a realistic enquiry into the legal controls on pollution one must examine critically the various statutory measures in force which relate to the environment. In some cases these reinforce the common law, in other cases they go further, creating new duties, breaches of which are offences punishable by fines and imprisonment.

Inland Water. At *common law* a landowner has a right to receive unpolluted water which he draws from wells on his land. One landowner has no more rights in respect of water flowing through his land than another. He simply has the right to use the water in a reasonable fashion and if his use of the water is unreasonable and to the detriment of his neighbour then the latter will have a right of action against him.

Common law therefore protects the rights of certain limited classes of people. However, the remedies of damages and injunction have failed to check pollution. This is a failing of the common law relating to all aspects of the environment. In the matter of inland waterways the *statutory* controls are found principally in the Rivers (Prevention of Pollution) Acts 1951-61.

The use of inland waterways is controlled by the river authorities. Their duties are to conserve, redistribute and augment water resources within their area, and to secure their proper use. All the river authorities are in turn responsible to the Ministry of Housing and Local Government.

River authorities are the key to a system of control of pollution of inland waters since it is only by their consent that local authorities and other bodies can discharge sewage and other effluents into streams. Local authorities are bound to receive domestic sewage and may agree to receive trade effluents, but can discharge into a stream only with the consent of a river authority. Where consent has been given no offence can be committed under the Rivers (Prevention of Pollution) Acts 1951-61, which make it an offence 'to cause or knowingly to permit to enter any stream any poisonous, noxious or polluting matter.'[2]

As has been noted, river authorities exercise statutory controls over discharges into watercourses. However, the growth of urban conurbations limits these powers. Domestic occupiers have a statutory right to discharge into local authority sewers and although a river authority can, by the imposition of reasonable conditions, control the quality of discharge into its rivers, in practice it cannot compel the local authority to comply immediately with its requirements. Therefore prior notice of any intended development affecting river waters is necessary if a river authority is to carry out its duties effectively. Unhappily, local authorities have all too often not given adequate co-operation to river authorities. Statutory obligation to give prior notice of intended developments to river authorities would be helpful.

Other problems arise with the difficulty of tracing pollution to its offending source, the rarity of prosecutions and the ludicrously inadequate penalties consequent on conviction. In 1968, for example, there were only 33 prosecutions. Maximum fines of £100 have little effect on large firms—in fact *actual* fines are ludicrously low, usually £5-£50 for pollution causing enormous damage.

The limited prohibition against polluting underground waters ('by means of a well, borehole or pipe') results in a river authority being helpless in many cases of pollution.

Finally, a major problem arises concerning the composition of the river authorities, and the question as to whom they are responsible. Representatives of the local authorities form a majority on the river authorities, therefore there is a conflict of interest between the desire of local authorities to dispose of the sewage and trade effluent which are their responsibility and the duty of the river authorities to exercise effective control over the quantity and quality of the pollutants they permit to be discharged into the water for which they are responsible. River authorities are not answerable to the electorate, and unfortunately pollution of inland waters is not, and is unlikely to be, an electoral issue.

Sea Fisheries. The Sea Fisheries Regulation Act (1966) established eleven sea fisheries districts, and local fisheries committees which are empowered to prevent by bye-laws the pollution of territorial water under their jurisdiction to the detriment of sea fish or sea fishing. However, these bye-laws are subordinated to the power of river authorities to grant consent to discharges and to statutory powers of local authorities to discharge sewage effluent. Thus sea fisheries committees are beset by difficulties. First there is pollution coming from beyond the territorial limit over which they have no control; secondly there is pollution from rivers, controlled by river authorities who have over-riding powers, thirdly there are discharges of sewage directly into the sea by local authorities over which the sea fisheries committee has no control.

Within these limitations, the committees themselves are open to criticism; there are those who interpret their powers narrowly, being prepared to act only when there is evidence of damage to fish or fishing interests. More commendable are those who consider any pollution of their area as affecting their interests and therefore to some extent adopt the role of pollution control authorities.

The discharges into coastal waters are many and various. No one authority has full knowledge as to what is discharged. Treatment of sewage discharged into the sea is almost universally inadequate, in some cases non-existent. Damage to the ecology of coastal waters is inevitable; therefore the sea fisheries committees must be strengthened and their jurisdictions extended.

Nuclear energy being of such awesome magnitude, it is not surprising that installations producing it come under strict control. The Department of Trade and Industry has control over the AEA; other government departments carrying out work in fields of nuclear energy are under the control of relevant Ministers. Anyone else wishing to build installations involving nuclear energy must obtain a licence from the Minister of Power. With the granting of a licence notice will normally be required to be given to interested authorities, i.e., any local authority, river authority, fishery committee, or other public authorities.

On granting the licence the Minister is bound to attach such conditions as appear to be necessary in the interests of safety. These conditions relate to the: (a) maintenance of devices to detect and record escapes of radiation; (b) preparations for dealing with an accident or other emergency on the site; (c) discharge of any substance on the site.

The Minister has appointed a Nuclear Installations Inspectorate. Inspectors may enter and inspect a licensed site, inspect documents and require information. The ultimate sanction is withdrawal of a site licence.

Regulations relating to the manufacture, handling and transportation of radioactive materials appear to be satisfactory in theory. The Radiological Protection Board, which replaces the AEA Health and Safety Division and the Radiological Protection Service, is independent in so far as it is responsible only to the Health Ministers. Its duties are mainly advisory. Monitoring will continue to be carried out by the Inspectorate.

Thus the combination of independent advice, strict control and opportunity for public criticism leaves little to be desired in theory. It remains to be seen how the system works in practice, bearing in mind that the industry has been subject to experiment throughout its growth. Constant scrutiny is vital in a field of operations where the margin of error is infinitesimally small and its consequences potentially catastrophic.

Air Pollution. This is probably the form of pollution most commonly noticed by the public. Enormous quantities of effluents originating from industrial plants, domestic fires, and motor vehicles are increasingly being discharged into the atmosphere, especially in urban areas. A distinction is made in the United Kingdom legislation between air pollution caused by chemical and other manufacturing plants and that produced by industrial or urban domestic premises. The former are governed by the Alkali Works Regulation Act, and the latter by the Clean Air Act (1956).

The Alkali Act is enforced by a small body of inspectors, less than 30 in 1967, who have the task of deciding what constitutes, for any given process, 'the best possible means' for preventing the discharge of noxious or offensive gases. The disastrous smog of 1952 exposed the inadequacy of previous control of atmospheric pollution and led to the implementation of the Clean Air Act, 1956. This provides the basis for the making of regulations for the control of atmospheric pollution by the Minister of Housing and Local Government. The Act prohibits the emission of 'dark smoke' from a chimney of any building. The only defences are that contravention was due to: (a) lighting up of a furnace which was cold and that all 'practicable' steps had been taken; (b) some failure in apparatus which could not have been foreseen or provided against; (c) the unobtainability of suitable fuel. The definition of 'practicable' in the Clean Air Act makes reference to the current state of technical knowledge and is an important clarification compared with that given in the Alkali Act. No new furnaces may be installed unless capable as far as is 'practicable' of being operated without emitting smoke.

An important feature of the Clean Air Act is that giving local authorities wide powers to make orders prohibiting the emission of smoke from the chimney of any building within specified smoke control areas.

Local authorities have different rates of progress in effecting smoke control areas. Some are reluctant to act because of the consequent charge on the rates. A great deal depends on the enthusiasm of the smoke control officer. Local opinion, for short-sighted reasons, may be against the imposition of control orders, and these feelings are often reflected by councillors.

Oil Pollution of the Sea. Control over pollution of the sea by oil is exercised under the Oil in Navigable Waters Acts (1955 and 1963). As far as UK territorial waters are concerned it is an offence to discharge oil, or a mixture containing oil, into them.

Further protection is afforded to the coasts and territorial waters of the UK by the existence of Prohibited Sea Areas. If oil is discharged from any UK ship within these areas an offence is committed. Further restrictions also apply to all British ships of over 20,000 tons, registered in the UK and built after May 1967. If a ship of this category discharges oil anywhere at sea an offence is committed.

The Ministry of Transport can require that UK-registered ships instal equipment to prevent oil pollution and may also direct harbour authorities to provide facilities for discharging oil residues.

This legislation and the regulations made under it stem from the International Convention for Oil Pollution Damage, and are complemented by corresponding legislation in the forty-one other signatory states. Evidence, however, shows that these measures have been largely ineffective. Spillage of oil per ton carried has been reduced but there has been a vast increase in the oil extracted and transported. Enforcement of the measures is very difficult, mainly as a result of the near-impossibility of tracing an oil-slick to its offending source. In 1968 there were no prosecutions by the Board of Trade, and only two in 1969. A 1969 amendment to the Convention seeks to allow a limited rate of discharge of oil per nautical mile which it is believed will not pollute sea waters. Large oil companies appear to have been co-operative in this respect but small companies and charter tankers present a greater difficulty. Close checks on the activity of all ships are necessary and methods of tracing oil discharges to the offending ship are required. The *Torrey Canyon* disaster of March 1967, which had catastrophic effects on birds, fish and flora, revealed the magnitude of the problem and also the lack of considered remedies for dealing with such a situation. The pollution of the sea by oil raises at the same time the problem of the pollution of waters by chemical detergents, pesticides, and so on. The detergent used against the oil on the English coast proved more poisonous than the substance it was meant to destroy. This same report concluded that the international arrangements in force for dealing with oil pollution were ineffective and insufficiently stringent. The exceptions to the prohibitions on the discharge of hydrocarbons were too great. The present limits of the prohibited zones (a hundred miles from the coast) were plainly inadequate. The report concluded that the slow speed with which legal concepts changed must be vigorously attacked, so that new ideas could be applied commensurate with the problems of the super-industrialised society of the twentieth century.

Solid Waste. UK legislation dealing with the dumping of solid wastes is clearly inadequate for dealing with the enormous growth of waste in a highly industrialised society. Most statutory provisions (to be found in the Public Health Acts and the Civic Amenities Act) were passed to deal with small accumulations of rubbish. As far as industrial waste is concerned, there are provisions for dealing with industrial alkali waste in the Alkali Act, 1906, but these have a narrow field of application.

The Public Health Acts empower local authorities to provide tips for the disposal of the household rubbish they collect. Trade refuse may also be collected, but charges must be imposed. Although local authorities are bound to dispose of waste, they will be liable if the manner of disposal they adopt proves to be a nuisance at law. The provision of tips is also subject to planning law. Local authorities also have the power to order the removal of accumulations of waste on private land. The Civic Amenities Act (1967) deals with the disposal of abandoned motor vehicles. The Alkali Act (1967) provides controls, in order to prevent a nuisance, on works in which acid is used on other substances capable of liberating sulphuretted hydrogen from alkali waste. The Act applies only to a small class of waste industrial products and requires only that no nuisance shall be created. Any legislation dealing with the deposit of waste products should concern itself not only with nuisance, but also with public health and amenity.

Noise. Local authorities take noise into account when dealing with applications for planning permission. However, restrictions relating to noise are rarely imposed, especially in the case of factories. Also, restrictions imposed at the planning stage obviously cannot be applied to premises built before planning permission was required.

There are also regulations in existence concerning motor vehicle noise but the police consider them unenforceable and rarely take action under them.

Legislation in the United States

US federal pollution legislation deals mainly with pollution of the water and air. The first water pollution legislation was passed in 1886, concerning New York Harbour. Air pollution legislation did not begin until 1955.

We shall deal with water pollution legislation first. Between 1899 and 1948 various Acts were passed prohibiting discharge of solid refuse into navigable waters, and prohibiting oil discharges from vessels into coastal navigable waters. In 1948 Congress took a national approach when it passed the first Water Pollution Control Act, which was extended in 1952 and 1956. This gave enforcement authority to the federal government if local efforts failed and included provisions for construction grants for waste treatment facilities. The Act was strengthened by the Water Quality Act of 1965 which set water quality standards for interstate waters. This Act also set up the Federal Water Pollution Control Administration (FWPCA).

The Nixon administration has legislation pending designed to strengthen the existing legislation. The existing Act required states to adopt quality criteria from the water quality standards set by the federal government and to draw up implementation plans for interstate waters by June 30, 1967. The new Bill extends the criteria and implementation required to *intra*-state and navigable waters. These have to be submitted one year after the Bill has been enacted. The existing Act makes no mention of effluent standards whereas the Bill establishes these and has provision for fines up to $10,000 per day for violation. The measure also empowers federal authority to investigate pollution and to require polluters to divulge the chemical nature of discharges. Provision is made for a $4 billion federal grant programme coupled with a $6 billion state and local programme for the construction of treatment plants.

Air pollution legislation began with the 1955 Air Pollution Control Act, which established a basic policy that primary responsibility for air pollution control rests at the local and state levels. The Clean Air Act of 1963, which replaced the 1955 law, called for financial assistance provisions for the development of state and local air pollution control.

An amendment in 1965 called for national standards for motor vehicle emissions. The Air Quality Act of 1967 was the last to reach the statute-book. It called for states to set air pollution standards on a regional basis and for regional standards to be enforced, locally if possible. The powers of local, state, and federal authorities were strengthened.

Under the Act the Department for Health, Education and Welfare (DHEW) must publish air quality criteria for those pollutants harmful to health, as well as reports on the technology that can be used to control those pollutants. The criteria are supposed to help states in developing air quality standards. The states had fifteen months in which to adopt standards and to adopt plans and schedules for implementation and enforcement of the standards. Their standards and plans must be submitted to DHEW for review and approval.

The Nixon administration has more legislative proposals under consideration for dealing with air pollution. These proposals demonstrate a feeling that the 1967 Act is a failure, particularly in relation to the long period of time before anything meaningful is achieved. The proposals would throw out the existing practice whereby criteria are issued governing air pollutants upon which states base the standards they adopt. Under the new proposal the criteria step is eliminated and full-blown standards are issued. The states would then have only six months to adopt implementation plans based on them.

The proposal would also authorise the National Air Pollution Control Administration (NAPCA) to establish national emission standards for 'hazardous' pollutants such as asbestos, beryllium and cadmium and for new sources which would endanger health.

As for enforcement procedures, the present Bill seeks to speed them up. At the moment the DHEW Secretary can ask the Attorney-General to bring suit in interstate pollution cases after 180 days' notice. In *intra*-state proceedings, such action can only be taken on a governor's request.

Under the present Bill, if a state failed to enforce standards, the DHEW Secretary could specify remedial action to be taken within 60 days. If the action were not taken, the Attorney-General could be requested to start injunctive action. The court could issue an order for remedial action and could assess a $10,000 per day fine beginning after the remedial period originally set by the Secretary.

But there has been for some time a growing dissatisfaction with existing and pending legislation. The initiative in the fight for the protection of the environment has had to be taken up by private pressure-groups in the courts.[3,4] One such pressure-group is the Environmental Defence Fund (EDF) which has been asserting in anti-DDT cases and other actions that people have a constitutional right to a clean environment. One radical theory by which the EDF is seeking to make the lawsuit a major weapon of conservationists, is that the Constitution's 9th Amendment (which says that the enumeration of certain rights elsewhere in the Constitution does not deny other rights retained by the people) can be invoked against polluters and others who disturb the environment unnecessarily. Most lawyers doubt whether the courts are ready to accept that argument. Most judges are extremely wary about venturing beyond precedent and known law and about deciding questions normally left for legislative determination, such questions as the general public's interest in clean air (as opposed to a mill's interest in cheaply disposing of its wastes).

Another radical theory discussed by US conservation lawyers was the 'public trust doctrine'. This holds that all land was once held in trust for the people by the sovereign (or government) and that the government cannot divest itself entirely of responsibility for the uses to which land is put, even though most of it has long since passed into private hands. The government must (according to the doctrine) see that no land, public or private, is abused or otherwise used in ways contrary to the public interest.

American conservation lawyers are hopeful that if courts should ever apply the trust doctrine or the 9th Amendment argument in a wide variety of environmental cases, this would force the executive and legislative branches to move at a faster pace in setting and enforcing standards for environmental protection. The Supreme Court's 1954 ruling against segregation in public schools triggered the release of dynamic social and political forces that produced the major civil rights legislation of the 1960s. However, on a pessimistic note, it should be remembered that the results of the civil rights legislation have failed to meet the hopes and expectations engendered by the campaign which led to its enactment. Is the same lack of concrete results likely to be the outcome of the present-day pressures for pollution control legislation?

It is, however, an indictment of the legislature and executive of the USA that such initiatives for environmental protection have to come through the courts, financed on a shoestring by the fund-raising efforts of local conservation groups who all too often will be fighting well-financed industries or government agencies. As was said above, the problem is too enormous to be left to private initiative. Administrations who neglect to act will find themselves damned in the eyes of future generations living in an environment where the life-support systems have been irretrievably altered.

Sweden

Sweden is nowadays considered to be the most progressive country as regards its approach to the pollution problem. The authorities there were roused to action as a result of the bitter public controversy over mercury pollution some years ago. The Swedish approach, it is now claimed, reflects foresight rather than an emergency response to an intolerable situation.

Sweden was one of the first nations to impose a total ban on the use of DDT, aldrin, dieldrin and other chlorinated hydrocarbons.[5] It has also undertaken a sizeable adult education programme aimed at creating in each community a corps of well-informed citizens who can organise public hearings and confront industrial and civil officials on what they are doing about pollution. In the year to October 1969, 250,000 persons received at least a few evenings' instruction on the technical and legal aspects of pollution. Eventually a thousand were picked, after further courses, to conduct public enquiries and in general agitate on behalf of pollution control.

The government has a special agency for dealing with pollution—the National Nature Conservancy Office, which exists as a branch of the Ministry of Agriculture. It has a staff of 175 and annual funds of $50 M.

Serving as the national policy-making body for anti-pollution activity is the Consulting Board for Environmental Problems, whose 24 members include 10 scientists, representatives of trade unions, industry, finance and the press.

Authority to deal with pollution problems is vested in local communities and regions, but the guiding force is the Conservancy Office with branches for nature conservation, water and air, and research. The guiding principles are to be found in stringent legislation which went into effect in July 1969, under the title of the Environment Protection Act. This contains provisions for the use of money ($50 M. over five years) to work against existing sources of pollution, and for the use of authority to prevent new pollution. Established industries will be given 25% of the cost of providing anti-pollution equipment. Industrial plants yet to be built will require permission from a specially created body, the Concessions Board for the Protection of the Environment, which has authority to set limits on how much pollutant may be put into the environment. Since one of the members of the Board will be drawn from the Nature Conservancy Office, a stronghold of anti-pollutant fervour, it is less likely that the board will become a tool of evasion, as has been the case with similar organisations in the United States.

Though the Swedes are a welcome example of foresight, will, and resources for dealing with pollution there are aspects of the problem that go beyond their reach. It is estimated that 15-50% of the sulphur dioxide in their atmosphere originates from other countries. Some of it comes from Britain, which has been looking after its own problems by introducing taller smoke-stacks so as to get the stuff carried off by winds.

This brings us to the international nature of environmental pollution and efforts made under International Law to bring control of the pollution of the environment within the sphere of international agreement.

International Law[6]

International legal controls on pollution are still at a primitive stage. The problems are difficult enough where a pollutant is confined to the boundaries of one country, but they are still more difficult where a pollutant spreads from one country to another. Not only is the problem of causation obviously aggravated by the difficulty of determining whether and to what extent pollution has arisen in another country, but there is no existing legal or administrative system which is able to deal with water pollution on an international scale.

Customary international law has scarcely advanced beyond the vague maxim taken from Roman Law, *Sic utere tuo ut non alienum laedas*—'use that which is yours in a manner which is not harmful to another'.

The chief areas of concern as far as international pollution is concerned are water and air. There are many treaties dealing with the use of waters, but most of them are imprecise, ineffective instruments and there is no uniformity between them. A lot of work has been done by various bodies, amongst them the Council of Europe, the Economic Commission for Europe, the International Law Association and the Institute of International Law, in formulating general principles governing the uses of international rivers, including the elimination of water pollution. But these general principles have not been implemented in any multi-lateral convention. The fact is that efforts to combat pollution on an international scale give rise to conflicts of interest between states and the application of the rule of law would not be completely pleasing to any interested state.

Pollution problems grew out of the spread of industrialisation in the nineteenth century, but notwithstanding early manifestations of pollution in Europe, these problems were largely untouched. In 1911 the Institute of International Law concluded that 'the exploitation of water for industrial, agricultural, and other purposes remains outside the provisions of the law'. Indeed, the primitive state of international law permitted the US Attorney-General to enunciate in 1895 the 'Harmon Doctrine', whereby he asserted that international law imposed no liability or obligations on the US concerning the use to which it put waters within its territory, regardless of how harmful such use might be to a neighbouring country (in this case, Mexico). However such a self-serving doctrine has not been widely relied upon, and in any case it is inconsistent with international legal principles, since the doctrine infringes the sovereignty of the lower riparian state in pleading the absolute sovereignty of the upper riparian state when in fact international law is based on the equality of states.

The few judicial decisions concerning international pollution tend to confirm the opinion that an injured state may take action against the state on whose territory the pollution originates.

In the Trail Smelter Arbitral Decision[7] concerning air pollution caused within the territory of the US by a foundry on Canadian territory, the tribunal concluded that: 'Under the principles of international law as well as the law of the US, no state has the right to use or permit the use of its territory in such a manner as to cause injury by fumes in or to the territory of another or the property or persons therein, when the case is of serious consequence and the injury is established by clear and convincing evidence.'

In the Lake Lanoux case concerning a dispute between France and Spain over the utilisation of the water from the lake, the arbitration tribunal pointed out that Spain's claims would have been upheld if the water returned to the lake by France after use had been of such a chemical composition, temperature or other quality as to damage Spanish interests. Failure by Spain to produce any evidence of injury barred the claim.[8]

Many international agreements have been made providing for the settlement of particular disputes, especially about the regulation and use of international rivers. These agreements underline the need for international co-operation and the urgency of developing more specific norms of international law by which disputes may be resolved. However, there are few treaties which establish comprehensive rules for the use of international environmental amenities.

Since the only principle of international law governing the environment would appear to be that of *'Sic utere tuo'*, which is too vague to be of general application, it is imperative that states must be prepared to limit the traditional concepts of state-sovereignty and submit themselves to be bound by multi-lateral treaties dealing with all aspects of environmental pollution. These treaties should be precise in the obligations they impose on the signatory states not to permit activities within their boundaries which would pollute the environment of a neighbouring state. Clear provisions should be made for agreed methods of arbitration in case of dispute, and each state should agree to be bound by the decisions of the tribunal both as to remedial action to remove the source of pollution and concerning penalties imposed by way of damages in cases where particular individuals have suffered loss as a result of pollution.

Man can no longer plan only on a national scale. The scope of industrial activity and size of the social problems to be solved in the world necessitate world planning and organisation. Pollution control within a single country is no longer possible.

PART FOUR

The Ultimate Solution — Social Reform or Revolution

20
Pollution and Political Priorities

In this chapter we are concerned with the complex interrelationship between social theory, social prophecy and social practice—a process formulated with clarity by Antonio Gramsci:

> 'In reality one can foresee only the struggle and not its concrete episodes; these must be the result of opposing forces in continuous movement, never reducible to fixed quantities, because in them quantity is always becoming quality. Really one "foresees" to the extent to which one acts, to which one makes a voluntary effort and so contributes concretely to creating the "foreseen" result.'[1]

As regards the question of pollution, all agree that it is a bad thing and should be stopped. But then most people agree that everyone should have an equal start in life, that the weak and helpless should be helped and looked after, that war is bad, that love is good, and so on. But still there is inequality, suffering, war and not enough love. So it is obvious therefore that the mere fact that everybody says that they are for a particular thing does not mean it will necessarily be achieved or obtained. Social life is after all a contradictory process; we have shown time and time again that the 'right' to pollute is advantageous to the polluter. Yet this polluter may well argue that he too is for a clean environment.

In view of the power of industry, in view of the enormous monetary advantage it may derive from violating our health and the environment, it might be asked how is it possible that any anti-pollution legislation ever gets on to the statute book at all? The reasons for controlling pollution are manifold. But the least powerful causal factor in achieving anti-pollution legislation has been the moral claim for a right to a clean environment. Industrialists and politicians are not necessarily devoid of intelligence and rationality. They are, however, subject to a great many conflicting pressures. The industrialist seeks to remain in business, if necessary at the expense of his fellows. The politicians and the state machine need to preserve a political and economic environment in which business can continue to flourish. To achieve this they may from time to time be forced not only to act against the working and middle classes but also to arbitrate in conflicts of interest between various ruling circles.

The state has access to expert committees which attempt to supply it with the latest scientific information and analysis of environmental problems. Generally, of course, little or nothing is done about such reports until a disaster occurs or a crisis develops. It did not, for example, become 'politically possible' to force through the Clean Air Act in England until after the terrible London smog killed several thousand people in 1952, a disaster that had been predicted twenty years before.

On other occasions, action occurs as a result of a conflict of interest, when one powerful group suffers by the actions of another. The Alkali Acts in England arose out of the conflict aroused between the landowners and the alkali manufacturers, who created the fumes. Remedial action is much slower where there is no such conflict within the ruling circles.

Contrast, for example, the difference in speed of response in Britain between regulating the conditions under which workmen used the bladder tumour-inducer beta-naphthylamine and the regulation of the conditions of use of insecticide-dressed seed. In the former case there was a fifteen-year delay between Hueper's announcement that he had shown experimentally that beta-naphthylamine caused bladder cancer in dogs, and the scheduling of bladder cancer as a prescribed industrial disease in 1952. In the latter case, two or three years of pigeon and game-bird kills by insecticide-dressed seeds sufficed to get action; in this case landowning interests were involved and the whole issue was hotly debated in the House of Lords. Of course it was right and proper that action was taken to restrict insecticide use, but the relative speed of action reflected the conflict of interest within the ruling circle. There is nothing new in this; it was noted by Marx in the nineteenth century that antagonism between landlords and industrialists aided the enactment of factory legislation. Whereas, the absence of such a conflict led to delays in producing legislation that would alleviate the atrocious conditions of work in mines.[2]

What is new in our time is that the industrial production process produces a wider range of compounds and products and has a higher level of interaction both between its component parts and the environment. There are thus more possibilities of dangerous pollution occurring. Whereas previously the traditional ruling-class conflict over pollution occurred between landowners and industrialists, it is now more complex. The sheer scale of modern industry has created shortages of what were formerly considered to be inexhaustible resources such as water. Thus the pollution of water by one industrialist may add to the production costs of another who requires clean water. The great advances made in analytical sciences have made it possible to show with increasing clarity the degree to which the environment is becoming polluted by an ever growing number of chemicals. The sheer magnitude of the problem produced by a hundred years of virtual neglect of environmental pollution has been massively documented. No government—and government is, after all, concerned with the future of society, if only to ensure the continuity of the business environment—can afford to ignore this completely.

There are also other processes at work which serve to put pressure on the government to make legislation, once enacted, ever more pervasive. One finds that legal control of pollution in one area forces increased pollution of another. Increased restrictions on fresh water pollution have led to more waste material being dumped in the sea. This will lead to pressure from the tourist and fishing industries to obtain restrictions on ocean dumping. Further, as soon as one industry finds itself restricted whilst others remain unfettered it begins to demand that all sections be equally restricted in the interests of 'fair competition'. A similar demand is raised for parity of anti-pollution legislation between different countries.

Politicians on Pollution

President Nixon in his Message on the Environment on February 10th, 1970, said:

> 'We in this century have too casually and too long abused our natural environment. The time has come when we can wait no longer to repair the damage already done, and to establish new criteria to guide us in the future.'[3]

Congress has enacted the National Environmental Policy Act, the purpose of which was 'To declare a national policy which will encourage productive and enjoyable harmony between man and his environment.'

To achieve these goals the President proposed that the federal government should spend $4 billion over a four-year period to control municipal, industrial and agricultural wastes, and that $6 billion be spent by state and local governments. In all, it was a programme which envisaged the spending of $10 billion.

He also promised to strengthen legislation and enforcement:

> 'We can no longer afford to consider air and water common property, free to be abused by anyone without regard to consequences... Instead, we should begin now to treat them as scarce resources... this requires comprehensive new regulations.'

Violation of the proposed national clean air standards could result in fines of up to $10,000 per day.

The opposition were quick to top President Nixon's proposals with a 50-point programme which contained a proposal to spend $25 billion on water pollution alone, and promised even stricter legislation.

In 1970 there was a successful effort to mobilise American youth into an environmental campaign. Many thousands of students were encouraged to celebrate Earth Day (22nd April, 1970). Commentators have pointed out that this particular 'student protest' showed an important qualitative difference from earlier protests; the unique experience of student protesters receiving the co-operation of authorities and industry. There had been no previous protests in which the authorities had banned traffic from certain streets to help the protest. In New York, both Fifth Avenue and 14th Street were closed to traffic. There was no previous precedent for such lavish aid in speakers and funds by private industry. In a book called *Ecology and Power* produced by an American radical group in 1970 it is suggested that government officials gave money and supplies to 'responsible' students who demonstrated for ecology and that businessmen provided cash for student conservation clubs and interested academics.[4] The whole idea of the Earth Day celebration is said to have originated with Senator Gaylord Nelson. Whilst there can be little doubt that many students found the environmental issue tied in with their other radical concerns such as the Vietnam War, poverty, race and the domination of the economy by the military-industrial complex, there is equally little doubt that the establishment succeeded in internalising their protest so that the environmental issue was presented in a manner which did not call the system into question. The people, government and big business were as one in opposition to pollution and the poisoning of the planet.

The idea that the pollution protest could be so taken over by the conservative politicians is not far-fetched. It is not, after all, a new political ploy. President Nixon has quoted President Theodore Roosevelt's view that 'the fundamental problem that underlies almost every other problem of our national life' is conservation and the proper use of natural resources.[5] His singling out of Theodore Roosevelt is not without significance. For Samuel Hays, in his book on the American Conservation Movement of 1890-1920, has explained that President Theodore Roosevelt looked upon conservation as the most important contribution of his administration in domestic affairs for 'it called forth patriotic sentiments which could override internal differences'.[6]

All over the world politicians vie with each other in producing anti-pollution rhetoric. Quite suddenly pollution has come out of political obscurity and become 'pollutics'. Pollution has become a political bandwagon which carries all parts of the political spectrum. Most politicians make the same speech.

The general arguments presented by the political establishment show concern tempered with an attempt to ensure that if the world doesn't turn into a nicer place during their term of office they have at least been able to pin the blame on the ordinary citizens' selfish desire for material possessions and children, neither of which were sacrificed for a clean environment.

Industrialists and Pollution

Controlling pollution is very expensive, but it is hard to find out exactly how overall figures for pollution control are arrived at. For what they are worth, here are some estimates. In 1966 US government sources estimated that pollution control would require about $60 billion over the next 15-20 years, about half of which would be wanted for industrial wastes.[7]

In water pollution control alone, sums in excess of $1,700 million were invested in each of the years 1967 and 1968 by public bodies. But American industry does not seem to be meeting its targets for pollution control investments. In 1968, the actual investments of $1,048 million compare very badly with the projected investments of $1,558 million. To realise the dangers in such a trend, if it is a trend, one has to remember the dynamic nature of waste treatments. New needs are being constantly created by replacement and growth. The costs of new needs developing in a year rose from $454 million in 1962 to $805 million in 1968. Furthermore, it is estimated that the United States entered 1970 with about $4.4 billion backlog of such needs. To eliminate this backlog over a five-year period will cost $10 billion because of the dynamic effect of growth, recapitalisation and price-level changes. Of course one could, by varying the annual investment, reduce or extend the period necessary to get rid of the backlog. But with any investment less than $1.5 billion a year it is mathematically impossible ever to eliminate the backlog of control needs; that is, the water pollution situation will keep on growing worse.[8]

Who is going to foot such enormous bills? Are the industrialists going to pay up or is the general public going to pay the costs? Jan Holdo, an executive of the Swedish-based Atlas Copco group, would prefer the latter course. He attacks what he terms the 'very popular and common' idea that industry should pay for the preservation of the environment. 'This is an easy way to get round the problem,' he said. 'Industry has no magic funds for the purpose.'[9] Holdo went on to add, 'The money and effort necessary for the preservation of the environment have to be taken from the users and consumers... industry is not the guilty party.' *Fortune* magazine expressed the same view in more sympathetic and sugary terms: 'No other group of Americans finds itself in quite as much of a quandary as the men who head the nation's largest corporations. The decisions they make, and the financial resources they commit, will be crucial to the clean-up effort. They are under great public pressure to act responsibly but on the other hand, they also have obligations to stockholders, employees, indeed to an economic system that thrives on ever-increasing production and profits.'[10] Some industrialists have sorrowfully explained how much it is going to cost to clean up. A representative of the National Steel Corporation, Pittsburg, claimed that '... to treat all its waters to standards recommended in 1965 by the Public Health Service, the steel industry would face a bill of $260M, or about one quarter of what its net earnings will total'.[11] This graphically illustrates the extent to which some industries are remaining profitable by passing on their costs to the public at large. If industries are able to make such enormous savings by violating the environment and damaging our health they are quite naturally very reluctant to pay the debt they owe to society and the environment. Another argument that industrialists frequently use is that stricter pollution control could make them less competitive in foreign markets. 'A country that allows itself to neglect the protection of the environment has the advantage of cheaper production compared with countries that care more about the environment. That means unfair competition.'[12] Does one detect here a feeling of envy for the more fortunate position of their Japanese counterparts? Japan has in part achieved its post-war 'economic miracle' by way of intense exploitation of worker and environment. The resulting pollution is so bad that it has been suggested that Japan be renamed 'the land of the rising scum'! Herbert P. Doan, then President of the Dow Chemical Co, has expressed similar feelings on the matter:

> 'Industry already lives in one world. We compete globally. The United States and all its problems will not do well unless industry is competitive. As a nation, we are not free to assign costs where we will. We are somewhat captive of economic practice in other parts of the world. Our society already assesses more costs to industry in the form of taxes than any other country by a wide margin ... It is only our enormous productivity that keeps us still competitive. To say this another way, if all countries put the burden of pollution costs on their industry, we could handle these costs too. If they don't—and they don't now and aren't likely to in the future ... then we can't either.'[13]

Such attitudes are both understandable and illuminating. It is understandable that in a 'competitive world' basing itself on naked self-interest, no one can afford to give his competitors even the slightest advantage. Further, such attitudes are illuminating to the extent to which one is able to see through the mystification produced by the continued use by these representatives of industry of 'we' and 'our', which is designed to make you and I, as ordinary citizens, identify ourselves with Dow Chemicals, ICI, Geigy, or Farbenfabriken, Bayer AG, according to whether we happen to be living in America, Britain, Switzerland or Germany. When Mr Doan talks of 'we' and 'our' he is projecting an idea that he (that is Dow Chemical) and we (that is, the ordinary citizen) are one. The ordinary people and Dow Chemicals, or any other corporation, are not an entity with common interests, 'we' do not control the policies of Dow Chemical nor are its profits 'ours'. The only thing that 'we' get free from industry that is 'ours' to keep is their muck and filth.

Let us now examine the great competitive world theory. Who are competing? Are the British people struggling against the

German people and the American people against the Japanese people? If they are, what are they competing for? After all, most people own very little, and could not survive on their own capital for even a year. The competition is in reality not between nations but between corporations which for historical reasons may be based in one country or another. The vast majority of people have in reality no country. This acceptance of the myth of the nation has long conditioned the peoples of the world to accept their subordination to a minority.

In the case of pollution the masses are being told that in the national interest they must either bear the pollution control costs incurred by industry or allow themselves and their children to be slowly poisoned and surrounded by ugliness. The industrialists' patriotic argument has to be exposed; as Dr Johnson said, 'patriotism is the last resort of scoundrels'. Who has ever heard of a private company being prepared to make losses in the national interest?

Whilst industrialists can no longer deny the existence of the pollution problem, they often claim it is exaggerated. Certainly such matters as inflation and 'law and order' interest them far more than pollution. In a *Fortune* poll, businessmen rated pollution as their number five worry, well behind Vietnam and inflation.[14] Another reason why businessmen show relatively less concern about pollution than these other matters is because the business community is better protected from environmental hazards than are the mass of people. Seventy per cent of business executives interviewed by *Fortune* magazine believed that their own and their family's health was not being directly affected by pollution. It is unlikely that executives will choose to live on industrial wastelands or polluted river banks or dirty their hands with irritant chemicals. But some of their unfortunate employees have less choice in the matter.

Fortune found, 'when it comes to remedial action for pollution, a dominant sentiment amongst businessmen is "caution". 'Caution' is a euphemism for doing little or nothing whilst pretending to be active. They are scared stiff that their profits would suffer—far better that mankind as a whole and the biosphere suffers.

Few, if any, corporations will control their pollution without the coercion of anti-pollution legislation. A firm which entered unilaterally into great expenditure on pollution control would place itself at a disadvantage with respect to similar firms that did not spend so much money. Legislation is necessary for collective action. As an executive stated in *Fortune*, 'In fairness to my stockholders... I can't make that first move.'

Under normal conditions the industrialists generally do not have to fear legislation for it will tend to err on the side that suits them best—'caution'. This has been the case since the British pioneered such legislation with the Alkali Act. *Fortune* found that over half the American industrialists interviewed spent less than 3% of their capital budgets on pollution control, and nearly all firms spent less than 10%. No doubt 'pollution control' was defined extremely broadly in the minds of the industrialists. Another expression of caution was noted in the view that changes should be made in processes used rather than products, for example, the automobile industry would prefer to fiddle with the internal combustion engine rather than find another power source. A majority of these firms complained that even their present 'cautious' expenditures were 'cutting into company earnings'.

When asked what the single most effective incentive would be to do more about the environment, the majority replied... tax credits. To quote Mr Doan again, 'The fact is, in spite of lots of talk and image advertising, and some progress—there is on all sides a tremendous amount of foot-dragging. It is a simple fact of life and... an economic necessity—that this foot-dragging could most easily, and at low cost, be turned into a foot-race by providing incentives to industry.' He added: 'We work well under the profit discipline. As far as I know, there is no evidence that we would do anything particularly well for the country, if we disregarded this discipline. We could make a real contribution, if we had the incentive; a paying customer.' Apart from the fact that it is precisely because they followed this profit 'discipline' that we suffer pollution, this sounds rather like: the public has let us dirty their environment whilst producing commodities— if they want clean air, water and soil, they'll have to buy it back! As Mr Holdo, the Swedish industrialist, puts it with admirable bluntness, 'If the customers and the consumers do not ask for and do not pay for silenced products and non-polluted air, industry cannot afford to solve the problem itself.'

It was a business magazine, not Karl Marx, which wrote, 'Industrial concerns exist not to serve people but to make profits. As this entails that the greatest return should be expected for the minimum investment, the attitude towards pollution control is that it is money spent without return.'[15] Let us leave the last word on the subject to the ex-president of Dow Chemicals.

'If assumptions about justifying profit are ideological, we will argue for ever about industry taking advantage of society. If assumptions are based on profit as a motivator, a useful social tool, an invention to put resources where we want them, then we should turn control of the environment to an advantageous opportunity for industry and the enterpreneur ... we will solve the problem relatively quickly and at the lowest cost to society.'[16]

Let us now examine how advertising has been used in relation to the call for greater control of industrial pollution.

Lord Cole, the former chairman of Unilever, is reported to have said that 'Advertising will have an increasingly important part to play in preventing the decay and abuse of our environment.'[17] He explained that advertising would also help to educate the public to distinguish between those activities which led to pollution and those which led to beneficial changes. By means of such educative advertising people would be prepared to accept innovations and not reject change out of hand.

Industry has indeed begun to use advertising to 'educate' the masses about the environment but not perhaps in the manner which Lord Cole would have had us believe. Ralph Nader in an interview said, 'General Motors has started a tremendous emphasis in advertising on its pollution control activities. This is all fraudulent.'[18]

Robert S. Leaf, president of Burson-Marsteller International, a public relations firm, said he would like to see more PR work on pollution: 'Many companies have excellent anti-pollution records, but they have never made it clear to the general public exactly what they have been doing to solve the problem.'[19] Not all companies using such an approach even bother to clean up. There is the now infamous story of the environmental advertising campaign of Potlatch Forests Inc. They produced an advertisement showing a beautiful clear blue river running through forested hills. The caption said, 'It cost us a bundle, but the Clear Water River still runs clear.' The *Wall Street Journal* were impressed enough to send a reporter to Lewiston, Idaho, where the company plant is housed, to have a look at this miracle of anti-pollution technology. The reporter soon discovered that the idyllic scene depicted in the advertisement was in fact photographed some fifty miles upstream of the plant. He also noted that the stream by the plant received forty tons of organic wastes each day and that the local air was also heavily polluted with SO_2 and smoke.

The crystal-clear stream advertisement is a popular motif amongst the lumber industry. One lumber firm placed that kind of advertisement in the magazines read by middle-class people to show how much it cared for conservation, but at the same time sent letters to its workers attacking conservationists because they were 'trying to limit the workers' rights to cut trees'.[20]

The magazine *International Management*, discussing the role of PR in pollution, says 'The public relations campaign not only aims to convince the public that industry is actively fighting pollution, but also to get across the message that waste disposal may mean higher prices.'

Such approaches seem designed to persuade people against pressing for major environmental improvements by frightening them with the spectre of inflation. They imply, 'look how much we are doing, we could do more but would not like to put up your cost of living. So be "reasonable" in your demands.'

Advertising and marketing generally use far larger budgets than industrial research. For example the Interpublic group, the world's largest advertising agency, had total billings in 1967 of over $721,000,000 from its clients. A single corporation, Unilever, spent £129,000,000 in the same year, advertising its products. The corporation who ask where they are going to find the money for anti-pollution control might care to dip into their advertising budget. After all, a great deal of pollution results from quite unnecessary products foisted on the public by carefully-designed advertising campaigns.

As we learn more about the dangers and extent of pollution,

new products have to undergo more and more searching safety tests. Manufacturers resent such costs, which do not lead directly to profit. A recent editorial in the scientific journal *Nature* claimed that '... it is anomalous that most governments—certainly the British and American—should expect manufacturers to pay for the costs of the tests necessary to prove that a new chemical can be safely used. So is there not a case for asking that some part of the increasingly onerous cost of testing should be borne by the central government ... [to make] sure that the inventiveness of the chemical industry is not stifled to a degree that will limit its social value.'[21] Surely the writer misses the whole point of private industry, which is to make profits for its owners. If they cannot make their profits without poisoning the public and the environment why should the public subsidise them? There are many things of obvious social value that are not done by industry because no profit can be anticipated.

A production controller for Carless Capel, the British hydrocarbon refining company, was reported by *International Management* as saying, 'We used to have a subsidiary company to turn sulphuric acid into copper sulphate for metal cleaning and other purposes, but the operation became uneconomic. Ideally we should like to be able to recycle but unfortunately it's cheaper to dump.'

When the attorney-general of California attempted to take action against offshore oil drilling he found most experts were financially tied to the oil industry. He is reported to have said, 'The university experts are afraid that if they assist in our case on behalf of the people of California, they will lose their oil industry grants.' One professor of engineering refused to testify 'because my work depends on good relations with the oil industry. My interest is serving the oil industry.'

Not only technical experts fear the strength of big business. Japan's Prime Minister, Mr Sato, is reported to have admitted, on Japanese TV, that whilst the powerful men of industry had so much influence on the government, it was difficult to see how effective anti-pollution legislation could be framed.[22]

It is clear that industry, and the politicians who for the most part represent their interests, are in a dilemma. They face a contradictory situation in which they are subjected to conflicting pressures. People are becoming increasingly less willing to put up with the filth and pollution with which they are forced to live, they protest and they may begin, if anti-pollution progress is too slow, to take direct action. Thus pollution could add fuel to the already burning social discontent in the advanced industrial societies. But in satisfying this growing demand for a clean environment many thousands of companies and even giant corporations could be financially threatened. To understand the complexity of this dilemma it would be useful to look at some of the political and economic factors which affect the kinds of decisions that are being and may be taken about pollution control.

Some Problems of Enforcement of Anti-Pollution Control

A government may express its willingness to improve the existing environmental standards but find that it is unable to enforce the new standards. The problem of enforcement of environmental legislation was discussed by the British Royal Commission on Environmental Pollution. That they chose to open their report by discussing this problem is significant, for the British officials have always maintained that their control was in advance of that of most other nations. For example, Mr F. E. Ireland, the chief inspector of the Alkali Inspectorate, said in 1970 that 'our achievements are believed to have surpassed those of any comparable industrial country'.[23] The statement of the Royal Commission on the question of enforcement is worth quoting in full:

> 'Legislation often fails in its purpose not on account of inadequate laws, nor through lack of technical knowledge, but because the laws are not being enforced, sometimes through indifference but usually because those responsible are unable (or unwilling) to meet the costs of controlling pollution. Some anti-pollution measures have been brought in only as a response to public anxiety over grave dangers (such as the London smog in 1952), rather than as part of a comprehensive policy for protecting the environment. Some kinds of pollution are likely to overtake present measures of control. The nation's resources for reducing

pollution are limited; difficult choices have to be made in their deployment.'[24]

Most of the reasons for not enforcing environmental legislation are economic. They range from corporations finding ways around the spirit of the law, from pollution-control workers striking for higher wages, to the ordinary householders unable to afford the cost of converting their heating to smokeless fuels, which often cost more.

One way to get round pollution control is to buy off complaints. Fluorine fumes are given off by brickworks, potteries, aluminium smelters and some cement works. Such fumes are very damaging to farm stock, causing a nasty condition called fluorosis. Fluorine fumes are particularly expensive to prevent and their control could, it is claimed, raise the price of bricks and aluminium by between 10% and 50%.

In Britain reported cases of fluorosis used to be fairly common but they have declined in recent years. It is suggested that a major factor in the decline has been an increased willingness for the farmers and brickworks to settle complaints privately, either by payment of compensation or by buying up affected land. The way the latter system works is that farms adjacent to brickworks are bought up by the brickworks and are then let to tenants whose leases stipulate that they are to avoid crops that could be affected by the fluorine fumes. In this way an industry can continue to pollute since nobody is complaining of financial loss. Thus control measures, instead of being preventive, are circumvented by buying off potential complainants.

Another approach is to play on the need for jobs and fears of unemployment. In this way a corporation might hope to blackmail local government officials into easing 'over-stringent' pollution controls by provoking a poverty-versus-pollution situation. This attitude is a traditional one. In 1828 a visitor to Sheffield, the centre of the British iron and steel industry, wrote that the people 'have become so accustomed to regard the increase in smoke as an indication of improved trade that they can see nothing in a clean sky but ruin'.[25]

In a recent case a German chemical company, BASF, wanted to build a $200 million petroleum plant at Hilton Head, South Carolina. The plant would have provided a thousand jobs for an area said to be 'one of the two poorest and job-hungriest districts in the whole US'. Eight miles from the proposed factory site was Hilton Head, an island resort patronised by millionaire executives. The local tourist industry, along with the local shrimp-fishing industry, led a very powerful campaign against the proposal for the factory. Since it is possible to build factories which do not pollute the environment, a rational person might think that there need not necessarily be any contradiction between a clean environment and availability of jobs. However, this view forgets the true reason for building and running petrochemical or any other plants—profits. When the tourist industry protested that the chemical plant would 'inevitably be a polluter' they were of course quite right. As Nicholas Faith of the *Sunday Times*[26] said, 'BASF, in effect, was only prepared to go to South Carolina on the basis that, as the local leader of the National Association for the Advancement of Coloured People put it, job opportunities were more important than cleanliness; or in other words it needed a free hand to do whatever is necessary to run a profitable dye-stuff and petroleum factory.' In the event BASF backed out; they had in any case taken over Wyandotte Chemical Company, whose Louisiana assets included a more suitable site for their plant. However, there is little doubt that such situations in which people are posed with the alternatives of poverty or pollution will become more frequent in the future.

To keep the environment clean and healthy requires workmen as well as machines. These men are often engaged in unpleasant work and quite rightly expect to be well paid. No environmental clean-up can be successful unless these vital workers are satisfied with their conditions. Yet it is quite obvious that they are often very displeased with their lot, for there have recently been many strikes by rubbish and garbage men and sewage-workers in many parts of the world. There are a number of very valuable lessons to be learned from the British sewage workers' strike in autumn 1970. Such events demonstrate the priorities of our society. The economic situation is hardly propitious for massive improvements in pollution control, for we are moving into a recession. If it comes to economic survival for business or a clean environment, who can doubt that in spite of the public demands for better pollution control business priorities will be put first?

21
Economic Rationality

We live in the most rational society that has ever existed, and yet ironically this is the first society whose actions threaten not only itself but the whole future of mankind. The reason for this apparent contradiction is the one-sided nature of our society's economic rationality. Before exploring this further, we must consider what is meant by rationality, or rational actions. Most of us feel sure of what we mean by an irrational act—for example, walking over a cliff as if a path continues out into space. A rational action therefore is one which corresponds to the reality of the situation. The better it corresponds to reality, the more rational it is. Obviously one cannot discuss rationality without bringing in knowledge of the situation in which one acts. For this reason some philosophers have thought it convenient to distinguish between factual and methodological rationality.[1]

Factual rationality is the idealised perfect situation, whereas methodological rationality is what we can normally hope to obtain. That is, one bases one's actions on what one understands to be the real situation. Of course, one is often wrong, not through irrationality, but through ignorance. As far as societies are concerned, one society is more rational than another in so far as it allows people, and encourages people, to develop a more complete knowledge of the situation in which they live and act. Capitalism, as we said in an earlier chapter, developed rationality further than any previous civilisation, because it developed the means of quantifying the results of its economic activities in terms of their profitability.

It will be apparent that any hindrance to the flow of knowledge within a society, or to the development of understanding, will make for less rationality and more irrationality. Here lies the key to the resolution of the contradiction mentioned earlier, as to why our society is apparently so rational and yet threatens the whole basis of life. Profitability, the source of its economic rationality, causes simultaneously the advancement of knowledge through research and education, and its hindrance through suppression and secrecy. Thus it comes about that the knowledge necessary to balance the economic equations used to determine the most rational course of action within our society is rarely available. There is of course another factor which determines the development and the utilisation of knowledge, and this is ideology, a topic we will examine in the next chapter.

It is the limitation placed by our society's social and economic structure on economic rationality that is the fundamental reason why our environment is polluted on such an enormous scale. For over a hundred years, manufacturers have passed on the costs of pollution to the population at large. Industry has never met the true costs of production, has never been sufficiently forced to restore or preserve the resources that it uses. This is also true of the unaccountable toll of human life shortened and wasted. Human life was equally used as just another cheap raw material suitable for exploitation. Fortunately human life is capable of fighting back and is not so easily destroyed. Thus capitalist economics seem incapable of assessing the full costs of production and even benefits are only seen in the light of crude monetary gain. To control pollution effectively it is necessary to develop a social accounting system which can assess and plan the costs and benefits of production according to real values based on life, not on the limited concept of profit.

Economists who have accepted the capitalist system of values see the situation from the viewpoint of an isolated, exploiting minority. They see the adverse effects which result from various processes, such as the spraying of crops with persistent insecticides, as essentially external. It is indeed seldom the decision-makers who suffer, nor always the consumer, but often a third party who has had no share in the benefits of the enterprise. The usual terms for such effects are 'externalities', 'spillovers', or 'social costs'.

The pollution problem is seen by these economists as one of 'spillovers'. 'Spillover' refers to unintended results of one's actions. They are further classified into 'external economies' (useful things) and 'external dis-economies' (harmful things). Given the complexities of the ecological processes of which man's activities form a part, the spillovers are usually external 'dis-economies' (often termed external costs).

The interesting thing about these externalities is that the person or persons responsible for creating them are gaining financially by not paying all the costs associated with their activities. Therefore they will only bear these external costs if forced to by special arrangements. For example: payment of damages, legislative effluent standards, subsidies, and taxes.

Cost Benefit Analysis

Within our society pollution results from a combination of four factors; an unwillingness on the part of polluters to take external costs into account, ignorance of the full implications of their actions, a tendency to base their decisions only on short-term considerations, and finally and most fundamentally the fact that within the present socio-economic structure it is far easier to obtain private economic rationality than social economic reality. Some economists hope that if these factors can be somehow analysed and taken into account, then economic cost-benefit models could be made that would take into account all the important ways in which man affects his environment; thus enabling more rational decisions to be taken. With such systems there would be no contradiction between economics and ecology, and pollution could be vastly reduced or eliminated, without damaging the economy. In theory, such economists are right, but we still have to face the problem of whether such schemes are really practicable within our existing social and economic structure. However, let us leave that question to one side for the moment, and look at the sorts of aids to decision-making offered by modern economics.

The most vaunted aid is called cost-benefit analysis. This attempts to find out how to achieve the lowest cost for any given level of effort, or the greatest effect for any given cost, and how to maximise the benefits and minimise the costs for the greatest total good.[2] Assuming we know the lowest costs of achieving various levels of pollution abatement, and the total damage costs of various pollution levels, we can construct models which show the economic effects of various levels of pollution control.

When such costs are plotted on a graph, it is sometimes found that the situations of no pollution control and total pollution control are equally costly; that one can have too much pollution control. According to such models, the economically optimal level of pollution control is the one in which the sum of the costs of pollution damage and the costs of pollution control is lowest. It might be, therefore, that further increases in the amount of control beyond a certain point cause this sum of damage and control costs to increase. In that case we would have carried pollution control beyond its most economically advantageous stage.

The trouble with such cost-benefit decision models or pollution 'damage functions' is that the data necessary to convert them from abstract models to the real-life situation are rarely available. Whilst it is possible to obtain reasonably accurate figures for the costs of control, the costs of damage by pollution are normally beyond assessment. Furthermore, the optimum level of pollution control would also vary according to the level of production and the profitability of various production levels. Profitability in turn would be affected by fluctuations in market prices. If we make further unlikely assumptions, such as perfect market competition and optimal distribution of income in society, we can build up a more dynamic model which can indicate these separate optimal controls for different outputs of pollution. Thus our scheme for identifying a single absolute level of pollution control collapses.

Many economists have struggled manfully to estimate the damages resulting from pollution in monetary terms. Some even come up with figures, many of which I have quoted in this book, but I would warn readers to take them merely as a guide. How can one estimate the costs of damage caused to health by air pollution? Also, the effects of pollution cannot be isolated from other factors, such as bad housing, damp climate, or smoking. Can damage to health truly be converted into monetary terms?

Lester B. Lave and Eugene P. Seskin made a brave attempt to measure the potential dollar benefits of air pollution abatement.[3] They concluded that a 50% reduction in air pollution levels in the major American urban areas would, in terms of decreased morbidity and mortality, lead to a net benefit of $2,080 million, or save 4.5% of the nation's illness costs. These monetary values were primarily based on what the person disabled or killed would have earned; but they believe that in terms of the values that people actually place on life, their figure underestimates the benefits. They point out that up to $15-20,000 a year is spent on keeping alive a person with kidney failure. We often hear today of workmates and neighbours clubbing together to get enough money to send a child with leukaemia, heart disease, or some other dreadful illness, to some special treatment centre. The sums expended bear no relationship to the 'economic' value of the child—they are based on love of somebody irreplaceable and priceless.

The task of converting external costs into monetary values has been approached in a number of ways. Some economists have suggested that where figures cannot be given for values, it might be possible to set some sort of constraint in the decision framework which would enable one to calculate possible quantifiable sacrifices. Such a situation nearly arose in Britain in connection with the pesticide/wildlife problem referred to earlier. It was calculated that the problem-causing organochlorine insecticides were worth £5,720,000 per year in increased produce to British farmers.[4] The cost to British farming of replacing these insecticides by suitable organo-phosphorus substitutes was estimated at £1,000,000 per annum. Thus one might have argued that the value of the peregrine falcon and other wildlife populations which were threatened by the continued use of these insecticides was £1,000,000 per year. But as Norman Moore has pointed out: 'Fortunately for conservationists there were also other reasons for banning Aldrin and Dieldrin.'[5] He was no doubt referring to the fact that much local damage had been done to game-bird populations. One knows shooting game in Britain is a sport of the élite, and a great fuss was made at the time by members of the House of Lords which played no small part in bringing about the ban.

Some economists believe that they can estimate the monetary value of pollution damage by examining changes in property prices. What this approach claims is that the market value of a house represents not only the house itself, but also its total environment; that the adverse effects of aircraft noise, for instance, on health, amenity and maintenance costs will be reflected in local property values. One objection raised against such an approach is that it assumes an abstract conception of perfect competition holds in the real-life property market, as though people have a perfect freedom of choice as to where they live.

Studies on the two London airports of Heathrow and Gatwick found that there were three principal factors that determined the depreciation of property values in the vicinity of the airports. They were: aircraft noise, general neighbourhood background noise, and the class of property. It was observed that increasing aircraft noise did indeed cause some property values to fall, but that such effects were less in those neighbourhoods that already had high levels of background noise. However, the most important factor of all was property class. For all levels of noise, the percentage depreciation in house value for 'high class' property was four times that for 'low class' property.

Consequently this method of quantifying damage has a built-in class bias against the poorer and working-class sections of the community. Since very small or zero values will be attached to their amenity loss they are therefore assumed to be unaffected by aircraft noise.[6] Another assumption in the property approach is that it assumes that people have been given full knowledge and understanding of the harmful effects to which they are being exposed. Furthermore, any damage effects may well be masked by the increased employment and trade that a new airport brings; it is quite conceivable that working-class property will hold its value because of the better local job opportunities. Is it objective to assume that because such people are too poor or too ignorant to move away from pollution, they are not suffering damage from it?

We have not exhausted the weakness of such models. So far, we have been assuming a situation in which isolated factories emit a single effluent, a known amount of which leads to a known amount of damage. Not a very common situation. The pollution unit generally puts out a mixture of different polluting effluents. The effects of these different pollutants cannot be isolated from the nature of the environment which they effect. They will do more or less damage in situation A than in situation B. They often become part of a general pollution produced by many factories and houses which have grown up together in a large sprawling mass without planning or foresight. Various components within this general pollution may react together to give greater or lesser degrees of damage than they would give in isolation. Consequently, ideal models would have to be able to deal with regional, national, and in some cases global units. Certainly it is now recognised that economic models for water pollution must be based on river basins, and air pollution models on conurbations. Such units rarely if ever correspond to administrative or national boundaries, and the accurate gathering of economic data under such conditions is thus made even harder. Neither does there exist an administrative body able to act on such data. Competing national states, where the gain of one is the loss of another, are incapable of creating a world organisation to deal with such world-wide problems. An organisation such as the United Nations in the present world situation either is impotent or becomes the tool of the most powerful states.

When faced with such complex tasks, how does the economist keep his sanity? How can any administrator accept with seriousness the figure which he presents? In reality, they don't—but save them as valuable ammunition against their opponents!

Nevertheless, these models cannot be dismissed as a complete waste of time. In attempting to make them, deficiencies in our knowledge, social institutions, and habits are highlighted. For instance, when attempting research into what goes into and out of factories it is the refusal of industrial concerns to cooperate in this matter which sabotages any attempt to assess real social costs and benefits. Although these attempts at rationality are to be admired, we are forced to the conclusion that under present conditions the search for a key to optimal pollution control is a modern variant of the search for the philosopher's stone.

Secrecy and Rationality

In order to plan one must know and understand exactly what it is one is trying to control. If vital pieces of information are withheld by private enterprises, then plans will be faulty however bright the planners. That is why planning is always limited and unsatisfactory under our present system. All economic rationality is restricted to private activity basing itself on the realisation of private ends, very often in conflict with the common good.

We live in an anarchic society, in which production is based on a multiplicity of separate corporations, each rationally pursuing its own separate goals. Pollution control, which would be social economic rationality, cannot be fitted easily into this scheme.

The pursuit of science and technological advances are the high points of human rationality. It is to them that we should be able to look for the final solution to the problem. Great technical and scientific skills need to be developed in response to the pollution problems. We must ask ourselves not only if they will be developed but whether they can be implemented. The relationship of industry to science and technology is a contradictory one, for it both develops and retards their development. The business enterprise is only interested in developing science and technology in directions which it believes to be profitable. 'There is one clear objective for research in industry: to create new ideas that lead to new products and new services... If research is to be successful in the efforts of industry it must be measured for profitability.'[7]

Giant corporations can to a large extent divert technological research into channels that are directly contradictory to the known facts of human needs. The cigarette industry, for example, conducts a never-ending search for obscure and futile data attempting to lessen the proven link between its product and disease. When an American federal report criticised the industry for not facing up to the health hazards or even admitting they exist the industry's PR machine protested that the report was a 'shockingly intemperate defamation of an industry which led the way in medical research to seek answers in the cigarette controversy'.

The giant corporations not only deform the content of much scientific and technical research, they also deform the capacity for clear objective thinking of their research workers. They are prevented from exercising their scientific integrity. An industrial scientist is not free to speak about his discoveries to other scientists without his employer's permission and when given permission to speak in public, it must only be to express agreement with the policies of the firm.

Secrecy is detrimental to social rationality. Scientific objectivity depends upon a process whereby the results of research are subject to the scrutiny and possible amendment of other workers in the field. Without a free flow of information this mechanism is damaged.

An instance of what can happen has been described by Professor Barry Commoner. A secret US government committee was set up to estimate the dangers of atomic fallout poisoning. Their estimate (later declassified), was an understatement by an order of magnitude. A major reason for the error was that the committee assumed that strontium 90 would enter plants only through the roots. Had a botanist been on the committee he could have informed them that many plants absorb nutrients through the leaves. Unfortunately there was no botanist present, and as the committee was secret nobody could tell them of this rather elementary (to a botanist) fact.[8]

It is a testimony to the strength of the ruling ideology that the fundamental principle of scientific universality has been eroded so easily by the necessity of business and national secrecy.

A report prepared for the American Association for the Advancement of Science[9] lists a new set of canons for applied scientists and technologists: 'Absolute secrecy where patents and industrial advantages are concerned; discretion in diplomatic matters where secrecy is essential during preliminary negotiations so that the negotiators are free to change their minds; security in matters of defence; confidentiality towards clients and patients; and loyalty to employing institutions where institutional aims are at stake.'[10]

The report makes no criticism of these 'canons'. On reading them one can see why no major revelations about the extent of pollution are made by private industry; or even mentioned by private industry. The closer one gets to a complete understanding of a situation the more rationally one can plan, but secrecy hinders the development and dissemination of understanding. Of course, in the short term, secrecy can be and is defended as a rational means to protect the narrow interests of a business enterprise against its rivals.

Business secrecy is yet another factor in the process whereby private economic rationality paves the way for socially irrational activity.

Any political campaign against pollution should seek to expose the true causes of pollution, stemming from the incompatability of the values of private enterprise with social rationality. It should demand that industry should have to establish the *safety* of their products. The evidence that they present to expert committees should be put on file for public examination.

Corporate executives, as well as corporations, should be held legally responsible for pollution and defective products.

All new technological agents should be put on a public register and not used commercially beyond a certain level until the general scientific community has had a good chance to study their toxicological and ecological effects. There should be a full right to *public* disclosure of business activities. For, as we have seen time and time again, whilst the immediate ends of technology are identifiable, their associated externalities are not so easily anticipated.

We should demand the opening of the books of capitalist enterprises. This should not only include financial accounts but a detailed account of their input and output of materials. Everything should be accounted for. The excuse of business secrets which enables industry to hide the extent to which it poisons the biosphere should be dismissed. In other words, we should demand the right to know what is happening and what is going to happen to us as a result of industrial activity. We should insist on an end to secrecy. It is in the struggle for such demands that conservationists will become conscious of the true nature of the system they are attempting to alter.

Hope for the Future?

In order to examine the chances of our society solving the environmental crisis we need to identify three different kinds of pollution problem: those in which the damage is *now* irreparable; those which are within the capacity of the system to avoid; and those which will occur because they are a result of the normal functioning of our social system.

First, there are certain kinds of damage and pollution which we will be unable to cure. Long-term pollutants such as strontium 90, radioactive carbon, DDT and heavy metals will remain for many years even if immediate steps are taken to prevent any more entering the environment. Once species of animals and plants and unique ecosystems, such as forests of giant redwoods, have been destroyed, they will never be recreated for they will have gone for ever. Lakes such as Lake Erie will require the investment of enormous amounts of labour and money to be cleaned up. It is hardly likely that such an investment will be made. Since there are so many competing needs, the best that can be hoped for is that no more effluent is allowed to enter them.

Secondly, it is quite likely that some forms of pollution will be greatly reduced, if not eliminated. A great deal of research and development has already been put into the problem of reducing aircraft noise, nuclear waste disposal, the reduction of city smoke, the cleaning up of factory effluent. In the advanced industrial countries we could hope for a cleaner environment, though not a clean environment, as far as such problems are concerned.

Finally, we can expect the continual emergence of new problems as well as the exacerbation of old ones through the working-out of certain processes inevitable within our particular type of economic structure.

Some will result from the uneven development of world economy. How can the under-developed countries afford to industrialise in a clean fashion if their better-endowed mentors find the implementation of anti-pollution measures so difficult? Already we can perceive the beginnings of a set of double standards—one for the industrialised nations and another for the Third World. DDT is now being slowly phased out of use in the advanced countries but in the Third World, where it is said to be essential for the maintenance of public health, it will continue to be used. Is it merely a case of DDT or malaria, as it is often presented? Alternative methods of controlling malaria and other insect vector-borne diseases could be developed. But their development would require the investment of trained manpower and money; only the advanced countries have the necessary resources. Very often the volume of technological exchange between the advanced countries and the under-developed leads to profits for the rich nations and new environmental problems, amongst others, for the under-developed nations. If in an advanced industrial nation individual industrial enterprises claim that they need some form of financial inducement to clean up; how much more could the Third World, struggling to develop modern industry, claim that without massive financial aid they will be driven to pollute in order to develop? Is it likely that the industrial nations of either West or East will provide such aid for anti-pollution control? As the Third World industrialises we can expect to see the old pattern of pollution, found from Manchester to Tokyo, repeated.

Within the advanced capitalist countries the business cycles and the vigorous but lopsided innovation process will exert a retarding effect on any efforts to establish a clean environment.

The business cycle of boom and recession acts on pollution control in a contradictory fashion. During periods of boom people are encouraged to consume even greater amounts of goods and energy. The more affluent one is the more, it seems, one wastes; as Thorstein Veblen once said, we live in a society characterised by an urge for 'conspicuous waste'. 'Keep up with the Joneses, buy, buy and buy even more', this is drummed into the mind day and night by subjecting people to a continuous barrage of advertising. In such an environment the expenditure of public funds on cleaning up the mess and waste left by orgies of consumption can barely keep up with the growth of waste. When we enter a period of recession the reverse happens; although consumption growth-rate drops we find that there is also less public money around for investment in new sewage schemes, that people—and industry—cannot afford to instal smokeless heaters and expensive anti-pollution equipment. In a period in which even cities declare themselves bankrupt, can one expect the cleaning-up of the environment to remain a major political interest?

Our society is an innovating society; new things, new processes, new fashions, and new patterns of behaviour are continually being developed and widely introduced. Vast sums of money are invested in social mechanisms designed to pass on 'the word' about new products, persuade people to buy them, and distribute them as rapidly as possible. This innovatory activity does not develop from the working-out in practice of some idealistic belief in a duty to supply people's needs. Rather it stems from the inner logic of a system based on producing commodities for exchange. Stress is laid on exchange, on sales and on profits; people are made to want and to desire, and therefore buy, and little or no consideration is given to the wider social implications of the introduction of a new product. America has, ironically, suffered more physical and material damage from its symbol of freedom—the automobile—than from any wars in which it fought this century.

These problems do not arise from any form of individual wickedness. They result from the actions of rational men operating within a particular set of social and economic constraints. In business there is no possibility of standing still, or sitting on one's laurels; one changes and grows or goes out of business either through a takeover or bankruptcy. As a consequence of such a ruthless struggle to survive, the private industrial enterprise continually strives to innovate, to steal a march on competitors and obtain a market advantage. All too often innovation reflects the market balance of power rather than any articulated public demand. We may expect new innovations to bring mankind new and unexpected environment problems, we may expect the history of the polluting innovations of the detergents industry to be repeated in new fields. We can also expect the military to add to their disgusting record macabre innovations which will help wreck the environment.

The struggle for survival of the business enterprise has another aspect besides that of innovation—its exact opposite, a search for stability by means of controlling society economically and politically. The giant corporations fear anything they cannot control; they need and seek a stable political and social environment within which to operate. Of course this is contradictory, for by many of their actions they continually send the system into instability. But nevertheless they attempt to control the political process. We can therefore expect the business world to attempt to retard and weaken any future legislation designed to clean up the environment if they feel it is a threat to their survival. The giant business corporations are not individuals but institutions whose roots ramify throughout the whole economy. To bring to heel an industry such as the auto-industry, responsible for so much pollution, would seem to be well-nigh impossible within the present social order. Contemporary America is the automobile; try to imagine it without the auto-industry and one sees something quite different. The top half-dozen corporations of the world belong either to the automobile industry or to the closely-related petroleum industry. Together they form a financial bloc without equal. Can one speak of democratic decision in relation to such colossi?

In the light of such possibilities one can expect increasing numbers of anti-pollutionists and conservationists to proclaim that their goal can only be achieved by far-reaching political changes rather than the present perpetual tinkering with the existing system. It is only because of this newly awakening consciousness that we can look to the future with any hope.

22
Pollution, Romanticism and Ideology

The protagonists in the pollution debate can be classified as hopeful Micawbers and gloomy Jeremiahs. Thus, when one asks what are the chances of survival in the future, one has the choice of an optimistic answer from some dreary unimaginative technocrat or an unbearably pessimistic reply from a visionary biologist. We have already devoted sufficient space to examining the crudities of the technocratic optimists, in particular we have stressed their intimate involvement with the industrial status quo. Therefore in this chapter I intend to examine the case presented by the 'gloom-mongers'.

It is first necessary to reiterate briefly some of the warning signs that the life-support systems of the planet are already under great strain. These are:

The increasing levels of radiation

the changing composition of the atmosphere
the elevated urban death and disease rate
the increasing incidence of respiratory disease, cancer and mental illness
the global contamination by chemicals such as DDT and PCB and by heavy metals like lead and mercury
the sudden reductions in the numbers of certain species
peculiar alterations in animal reproductive behaviour
the damage and destruction of enormous areas of fresh water like Lake Erie
the reduction of soil fertility
the contamination of the marine environment

Given such information, a Jeremiah-like attitude is understandable. Unfortunately biologists' political analysis rarely reaches the same level as their often excellent biological analysis.

If one examines their proposals for environmental reform, which may in theory be most thorough, one finds that they always imply that they could be carried out without fundamentally altering the existing social relations of production. In this chapter I shall concentrate on four aspects of such criticism: its moralising, its demands for a no-growth economy, an obsession with population, and romanticism.

Preaching Pollution Control

The moralistic approach to pollution begins by appeals to our 'sacred' and 'absolute duty to leave things no worse than we find them whatever the costs... and that if we are to have a right perspective about material things we must see them within the perspective of eternity'.[1] The problem with this approach is that it merely appeals to man's better nature and neither postulates nor demonstrates any material force whereby the demand could be realised.

Why does mankind go on ignoring such 'civilised advice?' Wouldn't most of the world accept in principle the view that life in all its forms should be safeguarded? The fact that this principle often fails to be followed means either that it is subordinate to some other fundamental principle or that civilised man is not what he thinks he is.

The De-Growth School

We discussed in the last chapter the material causes that drive the business enterprise into its continual struggle for profit and growth and stressed that this was not a result of some psychological kink of avarice.

The de-growth school believe that economic growth, that is, the continual creation of more wealth, is an 'economic shibboleth'. Not only is it not necessary for the system, but it is positively harmful. It is responsible for increasing population, increasing consumption, and increasing waste. All of which lead to an intensification of the pollution problem. Therefore, they argue, if we cut back on growth we shall ease the pressure on the environment.

But the de-growth school is faced with the problem that the economies they wish to control are prone to continuous inflation; that this inflation is exacerbated by continuous demands for higher wages; that the demand for more wages is a result of rising expectations, the natural desire of the working class to consume more of what their labour creates. Further, that this rise of expectations is growing more rapidly than the economy. This in turn puts the government into a contradictory situation where to check the inflation it must contract the economy, but to check the social unrest created by wage demands it needs to expand it.

The orthodox economists believe that only by continuous growth can a total collapse of the economy be prevented. History backs up their case; only disastrous economic crisis stops growth. The socially-conscious ones also argue that economic growth is the only way to obtain a more equitable distribution of incomes, but there is no evidence to show that this is true within the existing economic system. Despite continuous economic growth in the post-war boom distribution of wealth has altered very little.

In such a situation, to argue that growth creates our problems is incorrect. What we have is growth in the wrong places. To take an obvious example, there are far too many automobiles but not enough efficient public transport systems to create a viable alternative to the automobile.

What exactly do the exponents of 'no-growth' or reduced growth promise? Well, they claim that such a policy would allow a major transfer of resources from consumption and investment-for-growth into anti-pollution restorative and conservation measures. What they are not so clear about is how this 'transfer of resources' would be accomplished, or how the mass of the population, who are *under-* rather than over-privileged, will be persuaded to accept their static lot. Neither do they indicate how they propose to persuade the affluent ten per cent of the population, who own eighty per cent of the wealth, to redistribute it evenly amongst their less fortunate fellow citizens.

The prospect that would face most workers in a no-growth or 'steady state' economy in our non-egalitarian society would be slow promotion, poor prospects, and precarious jobs. Wages would be fixed. There would be no improvement in conditions. To maintain such a system would require more than changed attitudes, it would require a dictatorship.

Over-population

The mass media regale us continually about the world's population problem, that there are too many people, that there is not enough to go around, that pollution results from increasing population. If only, we are told, people adopted birth control our problems would be over.

This idea is a very old one and was popularised by Thomas Malthus in *An Essay on the Principle of Population* first published in 1798. Basically Malthus's theory is a very simple one. The sorry state of mankind is due to the fact that human populations tend to increase more rapidly than their means of subsistence. Malthus maintained that population increases geometrically whereas food supplies only increase arithmetically.

Quite naturally such a prospect leads to a feeling of unrelieved gloom.

'To remove the wants of the lower classes of society is indeed an arduous task. The truth is that the pressure of distress on this part of a community is an evil so deeply seated that no human ingenuity can reach it.'

'To prevent the recurrence of misery, is alas! beyond the power of man.'

In a world in which the wealth is unevenly divided, such a theory is extremely convenient for the privileged élites on two counts. Firstly, it shows that the situation is a result of 'natural laws' and therefore no one is to blame, and secondly, that very little or nothing can be done radically to alter the situation.

'... No possible form of society could prevent the almost constant action of misery upon a great part of mankind, if in a state of inequality, and upon all, if all were equal. The theory on which the truth of this position depends (is)... that population cannot increase without the means of subsistence... That population does invariably increase where there are means of subsistence... And that the superior power of population cannot be checked without producing misery or vice.'[2]

The basic weakness of Malthus' theory is that it is based on an over-simplified model of the relationship between soil and the human population. It fails to take into account the human ability to apply rational analysis to production, i.e. the development of science and technology. In fact Malthus was disproved by the first 150 years of the history of industrial capitalism, when the population of Europe grew threefold. During this period there was an unprecedented growth in the productive forces of mankind.

The 'gigantic inevitable famine' was prevented by technological innovations in agriculture, the opening of new lands, particularly in North America, and increased international trade.

The 'sickly seasons, epidemics, pestilence and plague' were virtually eliminated by advances in medicine and public health. The cultural changes accompanying such technical advances led to a fall in the birthrate.

Thus for 150 years the 'superior power of population' *was* checked without producing 'misery or vice' on the scale he predicted. However, since the end of World War II the ideas of Malthus have begun to gather more support, particularly over the last decade, when the hopes engendered by organisations such as the United Nations Food and Agricultural Organisation proved to be groundless. The gap between the developed countries and the underdeveloped was not closed—rather, it began to widen.

The usual explanation for this failure to develop the productive forces of the under-developed world is that a population explosion ate up the gains and advances. The population explosion itself was caused by a dramatic fall in death-rates without a corresponding fall in birth-rates.

Until relatively recently most underdeveloped countries had a 4% birthrate and a 4% death-rate. In such a situation the population remains fairly static. However, the situation can be radically altered by the following developments:

1. The introduction of efficient administration. This leads to an improvement in law and order, less internal warfare, better distribution of foodstuffs in time of famine. This can reduce the death-rate by as much as 1% in an under-developed country.
2. The introduction of public health measures to control epidemic diseases can reduce the death-rate by a further 1%. In Ceylon the use of DDT and an anti-malaria campaign reduced the death-rate from 2.2% to 1.1% in a mere seven years from 1945 to 1952. Contrast this with the fact that a similar fall in the death-rate in England and Wales took seventy years.
3. The development of medical services, doctors, nursing and hospitals on a scale similar to that found in the developed countries could bring about a further 1% reduction in the death-rate.

The full implementation of such services could thus bring about a total fall in the death-rate of 3%. If the birthrate continued at its former 4% the rate of population growth would move from under 1% to 3%. It is claimed that already many underdeveloped countries are experiencing growth-rates of over 2% a year.[3]

The significance of such percentage growth-rates becomes clear when one considers how long it takes a particular percentage growth-rate to double the original population. A 1% increase, if continued, causes a population to double within seventy years, a 2% growth will cause a doubling in thirty-five years, and 3% in twenty-four years.

The implications of such doubling rates have caused much speculation. The population-doubling times of under-developed countries range from 20 to 35 years, whereas those of the developed countries tend to be within the 50-200-year range. The world average is about 35 years. The last doubling of the world population took 80 years, the previous one 200 years. In pre-capitalist periods the doubling period has been estimated at 1,000 years. Some commentators have claimed that the current growth-rates are going to prove Malthus right after all. 'The battle to feed all of humanity is over. The famines of the 1970s are upon us—and hundreds of millions more people are going to starve to death before this decade is out.'[4]

The solutions put forward by population theoreticians generally lay major emphasis on birth-control. All such schemes will fail. Birth-control merely treats a symptom and leaves the root causes untouched. It will even have little effect on the symptoms. Many of the countries where birth-control is hoped to be a solution still have large numbers of people subsisting within peasant economies possessing extended family systems, in which cousins as well as parents and siblings live together. People who would starve in a wage economy, unless kept alive by public relief schemes, are fed and kept by their relations. To ask such a peasant to reduce his family is to ask him to reduce his insurance for old age. No family planning scheme could exert a major change unless it were part of a more general social plan which offered alternative means of support for the aged.

Furthermore, in agricultural countries farmers need plenty of seasonal labour—what better source is there than unpaid family labour? Thus a large family is seen as an asset. It is no good claiming that a family will save money by restricting its size if it believes that the gains will be lost in payments to hired labour. Thus mere campaigns for birth-control are useless; they have to be part of planned economic growth and industrialisation.

I am not arguing against birth-control. On the contrary, I consider its availability as an extension of human rights, in particular those of women. But talk of over-population directs attention away from the true causes of problems. Birth-control in itself will solve none of them.

If one poses birth-control as a cure for social ills then one must answer the question, what is 'over-population'? At first sight the answer seems easy; it is when the population exceeds the available resources. But is the concept an absolute or natural law of nature, or is it a relative law which holds in certain situations and not in others?

In societies of nomadic shepherds one finds population densities of 40-100 per square mile; nomads with agriculture 200-300; with extensive agriculture 200-500; regions with intensive agriculture 2,000-4,000. In regions of India where irrigation makes multiple cropping possible over 10,000 people per square mile can be kept alive, and finally, in the metropolitan areas of industrial societies densities of over 15,000 per square mile are found.[5]

What these data illustrate is the fact that with the development of more advanced productive forces the capacity of areas to support human populations can be increased. So-called laws of population differ from one mode of production to another. Thus no concrete discussion of human populations can be made in the absence of an analysis of the mode of production, the state of productive forces and the ideological superstructure of society.

No doubt one can imagine a hypothetical situation in which, given our present level of scientific understanding, there would develop seemingly absolute physical and biological limits to human population. But we have not reached any such limit yet. If there is 'over-population' it is a result of a major social and economic weakness, the stagnation of the productive forces.

The situation is that there are available technologies which could enable non-industrial countries to industrialise and meet their internal production needs within a generation; agricultural techiques that could give massive increases in food production, preservation and distribution methods to reduce current high levels of food losses; pharmacological techniques to control fertility at will. What rubbish to say that they are not applied because of population pressure! 'Over-population' results because they are not applied. We should direct attention to the problem of why they are not applied. Furthermore, within the advanced countries we find whole industries working at only 80% capacity; stagnant economies; over-production of food. It is not 'over-population' that prevents this underutilised capacity being diverted as aid to the under-developed countries.

The roots of these problems lie in an economic and social structure which has proved incapable of a socially rational use of resources. Malthus and his contemporary followers do not explain the real causes of poverty, misery or pollution. They do, however, serve to justify the *status quo* in society by presenting existing conditions as a result of 'natural laws' and therefore in essence insoluble. That is why all their arguments about economics and politics reduce problems to the level of biology — to discuss them at the level of society would mean questioning the continued validity of the existing social and economic structure.

Romanticism

A common denominator running through the critiques that we have discussed so far has been lack of realism. This often takes the form of romanticism.

The romantic critiques of the system have certain characteristics in common. They often clearly expose contradictions in the system, and show the enormous dangers to which we are being subjected by current industrial practice. But generally they make no attempt to analyse or explain the sources, development and trends of these contradictions. They simply regard the rape of nature and man by the existing system as something 'unnatural' or a mistaken and stupid deviation from some ideal norm. The romantic critique takes the form of moralising, denunciation and advice on how to correct existing failings. Its major weakness arises from the failure to realise that the contradictions and social and economic deficiencies that they expose express not only stupidity and ignorance, but the real economic and political interests of groups of the population.

A further weakness is that some of the romantics fail to acknowledge the progress that industry has made possible. It is true that many of these critics admit that technology and science possess a contradictory nature — that though they are not harmful in themselves, they become so under the prevailing economic conditions. But is it enough merely to leave a critique at that point and conclude that modern technology and industry are slowly killing us? Surely we must indicate the other side of the contradiction, namely, that modern technology also positively contributes to human progress. To simply concentrate on the negative sides of modern technology can lead one to conclude that the only way out of the impasse is to slow down technical growth, achieve some kind of 'steady state' or even, in extreme cases, go 'back to nature'. Such views are really a hindrance to understanding and remedial action for not only do they present an impossible solution but they pose false alternatives. We are not faced with either slowing down or carrying on along the present dangerous road; we could advance to a collective society which would increase the productive forces in a socially rational manner. But to see the necessity for this involves a definite break with the ruling ideology of our social system.

Ideology

In order to understand what ideology is we must remember that men exist as members of societies, not as isolated individuals. Human consciousness develops as a result of, and within particular sets of social relations. The particular form social relations take, especially the social relations of the productive process, to a large measure structures the types of ideas that any particular society develops. In our own society the legal, scientific, artistic, political and religious ideas that emerge are affected by its class structure. This does not mean that our ideas are necessarily totally incorrect or deliberately falsified. Simply, that when these ideas are generalised, at various levels of abstraction and sophistication, they produce a world view or *ideology* which represents the general interests of a particular social class.

Scientific ideas, so far as they contain social ideas, may create stresses in the ruling ideology. Two important occasions on which this has happened were the Copernican Revolution in the sixteenth century and the Darwinian theory of evolution in the nineteenth century. Both these advances helped to undermine the ideological authority of the Christian Church. Similarly today ecological theory, with its emphasis on the oneness and the interrelatedness of life, undermines the authority of the giant corporations. The private rationality whereby they determine their operations has been shown scientifically to be socially and biologically irrational. Such scientific advances could be said to polarise the ideological struggle in modern society into a *conservative* and a *progressive* ideology; the conservative ideology which seeks to maintain the present social structure, and the progressive which looks for revolutionary change. The liberal critics form a *compromise* ideology which seeks revolution without a revolution. As the present ideological struggle heats up, as a result of the intensification of the general economic and social crisis of the system, it will be the compromise ideology which will be the first to break. Some of its proponents will move to the right, others to the left. The proportions that this split will take will in large measure depend on the relative strengths of the polarised wings. Already, as the scale of political and economic change necessary to cope with environmental degradation becomes clear, some 'conservationists' show signs of a swing towards some kind of élitist authoritarianism which will force such reforms as birth-control on the 'ignorant' masses. But others will turn to a revolutionary position and ally themselves with the revolutionary sections of society.

23
Hope for the Future

Our society is in a deep and growing crisis. Industrial and population growth are making existing environmental problems worse and creating new ones. At the same time the gap between rich and poor and between the advanced and under-developed countries continues to grow. The resulting political climate throughout the world is hardly likely to be sensitive to the urgency of the environmental crisis. A mere recitation of such facts is a dispiriting task often ending in a pessimistic submission to the situation. Yet the path towards the solution of these problems is clear.

We know that all life, including ourselves, is engaged in what Charles Darwin called a 'universal struggle for existence'. However, the ruling strata of society have turned this natural struggle for existence into a brawl conducted on a global scale. In place of this, society requires for its struggle to survive the science and artistry of the judo champion, who bends his opponents to his will without breaking their bones. Such skill depends on the understanding of ecological laws. These are:

> That all living things are interconnected through ecosystems—nothing stands on its own. Thus no activity can be allowed to focus exclusively on a single element of the ecosystem.
> That we are part of the biosphere which necessitates global planning for society.
> That large-scale habitat changes are generally irreversible.
> That biological diversity, the wide range of species and habitants, is a dynamic store of information coded at varying levels from the molecular to the ecosystematic. To wipe out a species or habitat is rather like illiterate barbarians burning down libraries because they cannot see any use for books.
> That diverse ecosystems, with a wealth of different species, tend to be more stable and less vulnerable to change than simple ecosystems, those with only a few species.
> That the energy flows and material cycles of ecosystems can be altered, within limits, to increase the flow to man. However, any definition of optimum flow must be in relation to the *long term* needs of man.

If our activities are to be based upon such principles they must be planned and co-ordinated. Furthermore, many of the desires and demands associated with the so-called 'affluent society' of the West will have to be modified. The demands for more and more material possessions cannot be satisfied in a socially responsible society. There are a billion families in the world and they cannot all run a family car!

The belief that man has to produce and consume on a gargantuan scale in order to be happy is socially conditioned. Man has not always fulfilled himself through things. Today we can no longer afford to. A rational society would ensure that the knowledge of basic ecological principles and their relevance to its life style is part of everyone's education. However, social goals imposed from above would be resented and not achieved; people must assess their needs for themselves in a democratic fashion on the basis of a full explanation and discussion of the various possibilities. From such discussions overall social priorities could be established intelligently. These would form the basis for a series of national and economic planning goals designed to overcome such threats to civilisation as: environmental pollution, war, famine, the collapse of social morale, bureaucracy backed up by military and police dictatorship.

The attainment of such goals will require very careful planning so as to ensure that tasks are carried out in a manner which makes the best possible use of resources. The educational and mass media institutions will need reconstructing and to be given the job of explaining to people what is really happening. People will then be able to assess more realistically their social activities. The mass media would have to bring experts and the public together. By having to formulate clearly the possible consequences and implications of alternative policies scientists and other experts could achieve a greater social responsibility than is possible at the present time.

A significant indicator of the bankruptcy of our own society is the fact that many young people, including scientists, are dominated by a growing fear of science. The exposure of the scale of social and environmental destruction has driven them into a deep pessimism characterised by an almost total lack of belief in the positive benefits of science and technology. They have sunk into a vaporous ideology in which scientific and technological advance is represented as something threatening and uncontrollable. The use of science and technology in a planned fashion to achieve socially rational ends is in fact absolutely essential. Without them there is no hope for civilisation or the environment.

Technology has already made available extremely useful tools for planners. For example, great advances are being made in mathematical and analogue model building. One can make a model of a system that is going to be altered and feed into the model various possible changes. The model can indicate the consequences of alterations before any actual action is taken within the real-life system. Already we have seen that the use of space-simulators can teach men to land a rocket on the moon without their having been in space before. There is no theoretical reason why we could not so simulate environmental changes. In this way we could learn without disasters.

Every advance in computing will bring a corresponding advance in planning potential. Thus we already possess the tools to determine the optimum use of our resources. The linking of human needs, politics, science and education within social planning will lead to growing confidence and ability to control our affairs accurately in a creative and responsible fashion.

The fact that man is part of the biosphere rules out the possibility of solving the environmental crisis by planning within a restrictive national framework. Likewise the whole structure of modern society makes international planning imperative. This would require some form of world government, or at the very least a world planning and coordinating authority. The organisational tools and personnel already exist in an impotent state—the United Nations Agencies. Like so much of contemporary society, the United Nations embodies in a distorted form concepts and institutions which correspond to real needs.

There are in the UNO the seeds of a world planning commission. To name but a few UNO organisations already operating in the environmental field:

> The Department of Economic and Social Affairs, in particular its Office for Science and Technology, Social Development Division, Centre for Housing, Building and Planning, Population Division and Resources and Transport Division. There are also regional economic commissions, e.g. for Asia and the Far East, Africa, Latin America etc.
> The International Labour Organisation, which can deal with pollution of the workers' environment.
> The Food and Agriculture Organisation.
> The United Nations Educational, Scientific and Cultural Organisation.
> The World Health Organisation.
> The International Maritime Consultative Organisation.
> The International Atomic Energy Agency.
> International Civil Aviation Organisations.

These organisations, or rather their revolutionised descendants, could play a crucial role in seeing that depollution is achieved on a global scale, in the most rational manner possible. Furthermore, in the political and cultural battle for clarity as to what goals and targets need to be set, they could fulfil a vital educational task.

Rational planning is the path out of our crisis but it is a path which will remain closed until the social and economic obstacles which block it are removed. The major obstacles are bureaucracy and the capitalist system of private property with its relentless drive for profit. Their removal is the necessary first step towards a decent and healthy world. Socially rational planning will be impossible until the working people control the means of production on a global scale. That is: industry, land, banks, transport and distribution will have to be nationalised and run according to the *collective* will by a global socialist society.

Towards a Healthy Environment

A socialist society, unlike a capitalist society, starts from social requirements based on the demands of the population. Members of such a society will claim as a legal and political right a healthy environment. Further, because of the political nature of such an aim great importance will be attached not only to orthodox medical criteria, but also to 'subjective' criteria. Consequently, because individuals have different 'tastes', it will become necessary to develop an environment in which the individual can exercise choice, an environment in which individuals can increasingly change its components in accordance with their needs and moods at any time. An environment which gives a man *freedom*, that does not constrict his will.

Ivan Pavlov thought that man had an inbuilt 'freedom reflex'. When this 'freedom reflex' is inhibited by anything that an individual holds to be *avoidable* or *unnecessary* he suffers physiological and psychic injury. That is why in contemporary society, even where there has been a quantitative increase in freedom, as a result of the rise in the level of productive forces, the youth feel unfree. The rise in the educational level has produced increasingly large numbers of people who see more and more of the present social, economic and environmental restrictions to be *unnecessary* and avoidable. Contemporary society has created yet another internal contradiction for itself —it raises expectations, to a level above that which can be provided for within the present socio-economic framework. When the 'freedom instinct' is inhibited men become uncomfortable and ready to raise their level of consciousness in order to seek out the means to smash the restrictions. This process has begun and the revolution and the environmental struggle will become one.

Notes

References are given only in brief. The full details of books and articles are contained in the bibliography.

CHAPTER 1
1 E. J. Kormondy, *Concepts of Ecology*, Prentice-Hall, 1969.
2 G. H. Hutchinson, 'The Biosphere', *Scientific American*, September 1970, pp. 45-55.
3 V. Kovda, 'Contemporary Scientific Concepts Relating to the Biosphere' in *Use and Conservation of the Biosphere*, UNESCO, 1970.
4 T. S. Kuhn, *The Structure of Scientific Revolutions*, Chicago, 1962.
5 F. C. Evans, *Science*, Vol. 123, 1956, p. 1127.
6 R. Glen, *Journal of Economic Entomology*, Vol. 47, 1954, pp. 398-405.
7 A. G. Tansley, *Ecology*, Vol. 16, No. 3, 1935.
8 Raymond L. Lindeman, *Ecology*, Vol. 23, No. 4, 1942, p. 400.

9 Hugo Sjors, *Svensk Botanisk Tidskrift*, Vol. 49, Nos. 1-2, 1955, p. 160.
10 For an excellent general review of these cycles, see *Scientific American* (Biosphere issue), op. cit.

CHAPTER 2
1 B. P. Uvarov, *Locusts and Grasshoppers*, London, 1928.
2 C. Elton, *Voles, Mice and Lemmings, Problems in Population Dynamics*, 1942.
3 J. B. Calhoun, *Journal of Mammalogy*, Vol. 33, 1952, pp. 139-59.
4 A. J. Milne, *Journal of Theoretical Biology*, Vol. 3, 1962, pp. 19-50.
5 If I have seemed to stress the element of change rather than harmony in the environment it is a result of attempting to redress the balance in the current debate on the environment.
6 M. Bates, 'The Human Ecosystem', in *Resources and Man*, W. H. Freeman & Co., 1969.
7 V. Kovda, 'Contemporary Scientific Concepts Relating to the Biosphere' in *Use and Conservation of the Biosphere*, UNESCO, 1970, p. 18.

CHAPTER 3
1 F. Engels, *The Dialectics of Nature*.
2 K. Marx, *Capital*, Vol. 1.
3 J. Steuart, *Principles of Political Economy*.
4 D. Dillard, *The Economics of John Maynard Keynes*.
5 G. Bloch, *Fourth International*, Vol. 6, No. 1, 1969, p. 9.
6 V. G. Childe, *Man Makes Himself*.
7 B. Malinowsky, *A Scientific Theory of Culture*.
8 R. B. Lee, 'Kung Bushman Subsistence: an input-output analysis'.

CHAPTER 4
1 V. G. Childe, *The Town Planning Review*, Vol. XXI, 1950, pp. 3-17.
2 E. Mandel, *Marxist Economic Theory*.
3 L. Mumford, 'City: Forms and Functions'.
4 G. Sjoberg, *The American Journal of Sociology*, Vol. LX, 1955.
5 E. Mandel, op. cit., p. 122.
6 J. Needham, 'Science and Society in East and West'.
7 K. Marx, *Capital*, Vol. 1.

CHAPTER 5

1 Abel Wolman, 'The Metabolism of Cities'.
2 R. J. Benthem, 'The Management of the Environment in Tomorrow's Europe', Council of Europe *Report on Urban Conglomerations*, 1970.
3 C. A. Doxiadis, *Impact of Science on Society*, XIX (2), 1969, pp. 179-93.
4 I. R. Passino, 'The Management of the Environment in Tomorrow's Europe', Council of Europe *Report on Industry*, 1970.
5 Composition of clean, dry air near sea level.

Gas	Concentration in parts per million
Nitrogen	780,900
Oxygen	209,400
Argon	9,300
Carbon dioxide	318
Neon	18
Helium	5.2
Methane	1.5
Krypton	1
Hydrogen	0.5
Nitrous oxide	0.25
Carbon monoxide	0.1
Xenon	0.08
Ozone	0.02
Nitrogen dioxide	0.001
Ammonia	0.01
Sulphur dioxide	0.0002

(From *Cleaning our Environment*, American Chemical Society, Washington, D.C., 1969, p. 24)

6 W. T. Russell, *Lancet*, Vol. 207, 1924, pp. 335-9.
7 J. R. Goldsmith, 'Effects of Air Pollution on Human Health', 1968.
8 Petrilli, European Conference on Air Pollution, Council of Europe, 1964.
9 L. M. Sabad, 'Air Pollution Control', WHO Seminar, 1967.
10 Lung Cancer Mortality rates for England and Wales 1950-1953. (The actual figures have been converted to a reference figure in which average mortality = 100. This figure is separate for each sex and therefore represents a higher number of cases for men than women.)

	Men	Women
Very large towns	125	121
Towns of more than 100,000 inhabitants	112	101
Towns of 50,000 to 100,000 inhabitants	93	88
Towns of less than 50,000 inhabitants	84	86
Rural areas	64	77

(Table taken from article by R. E. Waller, *British Journal of Cancer*, 1952, Vol. 6, pp. 8-21.)

11 A. J. Haagen-Smith and L. G. Wayne, article in A. C. Stern, *Air Pollution*, Vol. 1.

12 The ocean may be the largest known natural source of carbon monoxide, contributing possibly as much as 5% of the amount generated by the burning of fuels, J. W. Swinnerton et al., *Science*, Vol. 167, pp. 984-6.

13 H. E. Stokinger and D. L. Coffin, in Stern, op. cit.

14 Delebruyere, European Conference on Air Pollution, Council of Europe, 1964.

15 V. A. Rjazanov, 'Air Pollution Control', WHO Inter-regional Seminar, 1967.

16 *Cleaning Our Environment*, Report of the American Chemical Society, 1969, p. 37.

17 Stokinger and Coffin, in Stern, op. cit., p. 447.

18 B. Slonim and K. Estridge, *Journal of Environmental Health*, Vol. 31, No. 6, May-June 1969.

19 *The Protection of the Environment*, (Cmnd. 4373, 1970).

20 T. S. Chow and J. L. Earl, *Science*, Vol. 169, pp. 577-80, 1970. Report finding up to 300 micrograms/litre of lead in rain-water.

21 D. Bryce-Smith, *Chemistry in Britain*, Vol. 7, No. 2, February 1971, p. 55.

22 M. W. Oberle, *Science*, Vol. 165, 1969, p. 991.

23 *Fresh Water Pollution Control*, Council of Europe, 1968.

24 H. M. Ellis, *Solid Waste Disposal*, Council of Europe European Committee for the Conservation of Nature and Natural Resources, Strasbourg, 1967, p. 1.

25 In *Cities*, A Scientific American Book, 1965.

26 *Environmental Quality*. Council on Environmental Quality, 1970.

27 I. Abey-Wickrama et al., *Lancet*, 13 December 1969, pp. 1275-7.

28 'A Review of Road Traffic Noise', Road Research Laboratory, 1970.

29 N. Lee, 'Studies in Environmental Pollution' (Paper 8), Pollution Research Unit, University of Manchester, 1970.

CHAPTER 6

1 Poison in our Air, Conference on Air Pollution, United Steel Workers of America, 1969.
2 K. Marx, *Capital*, Vol. 1.
3 I. Gough, *Inequality in Britain*, unpublished report, University of Manchester.
4 J. E. Lunn et al., *British Journal of Preventive and Social Medicine*, Vol. 21, 1967, p. 7.
5 J. R. T. Colley and D. D. Reid, *British Medical Journal*, Vol. 2, 1970, pp. 213-17.
6 P. Gregory, *Polluted Homes*, 1965.
7 J. N. Roehm et al., *Chemical and Engineering News*, 29 June 1970.
8 M. G. Jones and J. T. Davies, Special Report No. 61, Iron and Steel Institute, London 1958.
9 I. J. Selikoff, 'Community Effects of Pneumoconiotic Dusts', *Pneumoconiosis: a symposium*, WHO, 1967.
10 *The Prevention and Control of Dust Diseases in Industry*, British Chemical Industry Safety Council, London.
11 R. S. F. Schilling et al., *British Journal of Industrial Medicine*, Vol. 12, 1955. Gibson and C. B. McKerrow, 'Vegetable Dusts as an Industrial Hazard', *Journal of Pakistan Medical Association*, 1968.
12 Senior Medical Inspector Advisory Panel, 1967.
13 S. R. and O. 1931, 1140.
14 Senior Medical Inspector Advisory Panel, 1967.
15 G. W. Wright, *American Review of Respiratory Disease*, Vol. 100, No. 4, 1969.
16 T. S. Scott, *The Practitioner*, Vol. 195, 1964, pp. 535-43.
17 D. O'B. Hourihane, *Thorax*, Vol. 19, 1964, p. 268.
18 Selikoff et al. cited by G. W. Wright.
19 T. S. Scott, *Occupational Health*, Vol. XV, No. 6, 1963.
20 R. A. Case, *Annals of Royal College of Surgeons of England*, October 1966.
21 C. Veys, *Journal of the National Cancer Institute*, 1969, p. 48.
22 J. M. Davies, *Lancet* ii, 1965, p. 143.
23 R. Doll et al., *British Journal of Industrial Medicine*, Vol. 22, 1965, p. 1.
24 W. C. Hueper, *Occupational and Environmental Cancers of the Urinary System*, 1969.
25 H. W. Mitchell, *American Journal of Public Health*, Vol. 55, No. 7, pp. 10-15.

26 D. Walder, 'Decompression Sickness in Tunnel Workers', in C. N. Davies, R. Davis and F. Tyrer, *Abnormal Physical Conditions at Work*, 1967.
27 A. Glorig, *American Journal of Public Health*, Vol. 51, 1961, pp. 13-38.
28 A. Bell, *Noise*, WHO, 1966.
29 M. L. H. Flindt, *Lancet*, 14 June 1969, pp. 1177-81.
30 *Vibration Syndrome*, Interim report by the Industrial Injuries Advisory Council, London, HMSO, Cmnd. 4430, June 1970.
31 There was a minority report which said: 'In our view the evidence which we have received confirms that there is a hard core of cases in which vibration-induced White Fingers results in change of job, sickness absence, loss of earnings, social handicap and pain.' Ibid. p. 14.

CHAPTER 7
1 M. E. Maldague, Agriculture and Forests, Council of Europe, European Conservation Conference, February 1970.
2 C. E. Yarwood, *Science*, Vol. 168, 1970, pp. 218-20.
3 *Modern Farming and the Soil*, HMSO, 1970.
4 H. Walters, *Journal of the Soil Association*, July 1970, pp. 149-70.
5 H. Walters, *Journal of the Soil Association*, April 1970, pp. 113-15.
6 *Chemical and Engineering News*, 7 September 1970.
7 R. A. E. Galley and J. G. R. Stevens, *Span*, Vol. 9, No. 2, 1966, pp. 107-9.
8 *Journal of Economic Entomology*, Vol. 40, 1947, p. 49.
9 G. Löfroth, 'Man and DDT Compounds', *Deutsch-Schwedisches Symposium auf Gebilt der Umwesthygiene*, Baden-Baden, 1969.
10 A. H. Conney, *Pharmacology Review*, Vol. 19, 1967, pp. 317-66.
11 C. F. Wurster, *Chemical Fallout: Current Research on Persistent Pesticides*, 1969.
12 Review of the persistent organochlorine insecticides, HMSO, 1964.
13 US Congress: Committee on Government Operations: Sub-Committee on Re-organisation and International Organisations: Inter-agency Coordination in Environmental Hazards, Pesticides, Washington, DC, 1963-4. (The 'Ribicoff Committee Hearings'.)
14 F. E. Egler, *Bioscience*, Vol. 14, No. ii, 1964, pp. 29-36.
15 C. M. A. Baker et al., *Comparative Biochemistry and Physiology*, Vol. 17, 1966, pp. 467-96.
16 L. J. Carter, *Science*, Vol. 163, 1969, pp. 548-51.

17 S. Novick, *Environment*, Vol. 12, No. 8, 1970, p. 25.
18 *The Observer*, 1 November 1970, p. 1.
19 G. Löfroth, *Naturwissenschaften*, Vol. 57, No. 8, 1970, p. 393.
20 *New Scientist*, 29 October 1970, p. 206.
21 'Deficiencies in Administration of Federal Insecticide, Fungicide and Rodenticide Act', Hearings before a Sub-Committee of the Committee on Government Operations, Washington, June 1969.
22 R. van den Bosch, *Environment*, Vol. 12, No. 3, 1970, p. 219.
23 R. F. Smith and R. van den Bosch in *Pest Control* ed. Kilgore and Doutt, 1967.
24 E. F. Knipling, *Journal of Economic Entomology*, Vol. 55, 1962, pp. 782-6.
25 L. D. Christendon, *Advances in Chemistry* series, Vol. 41, 1963, pp. 31-5.
26 M. Jacobson, *Journal of Organic Chemistry*, Vol. 25, 1960, p. 2074.
27 P. DeBach (ed), *Biological Control of Insect Pests and Weeds*, 1964.
28 F. J. Simmonds, *Journal of the Royal Society of the Arts*, Vol. 115, 1967, p. 880.

CHAPTER 8
1 *Fresh Water Pollution Control*, Council of Europe, 1966.
2 UN Doc. E/Conf. 39/1, Geneva 1963.
3 FAO Symposium on the Nature and Extent of Water Pollution Problems Affecting Inland Fisheries in Europe, 1970.
4 WHO Technical Report No. 318, 1966.
5 Dr Key, 4th European Seminar for Sanitary Engineers, Opatija, Yugoslavia, 1954.
6 P. Vivier, *La Pollution des Cours d'eau*, Académie d'Agriculture de France, Extrait du procès verbal de la séance du 25 juin 1958.
7 UN Doc. E/Conf. 39/1, Geneva, 1963.
8 *Clean Water for the 1970s*, Federal Water Quality Administration, June 1970.
9 B. Commoner, 'Balance of Nature' in *Providing Quality Environment in Our Communities*, ed. W. W. Konkle, US Dept. of Agriculture, Washington DC, 1968.
10 *Chemical and Engineering News*, 19 May 1969, p. 43.
11 *Nato Letter*, Vol. XVIII, No. 2, 1970.
12 *Chemical and Engineering News*, 14 October 1968, p. 21.
13 E. J. Kormondy, *Concepts of Ecology*, 1969.
14 *Chemical and Engineering News*, 22 June 1970, p. 36.
15 Report on an Inter-Regional Seminar on Water Pollution Control, WHO, New Delhi, India, November 1967.

CHAPTER 9
1 The Mercury Pollution Problem in Michigan and the Lower Great Lakes Area, Michigan Water Resources Commission, May 1970, p. 4.
2 G. Löfroth, *Methylmercury*, Swedish Natural Science Research Council, 1970.
3 *Chemical and Engineering News*, 20 July, 1970, pp. 18-19.
4 *Chemical and Engineering News*, 27 July 1970, p. 86.
5 G. Löfroth, Report on Methylmercury (prepared for WHO) Stockholm, 1968.
6 *Chemical and Engineering News*, 22 June 1970, p. 37.
7 *Chemical and Engineering News*, 13 July 1970, p. 15.
8 A. L. Hammond, *Science*, Vol. 171, 1971, pp. 788-9.
9 *Chemical and Engineering News*, 10 August 1970, p. 16.
10 C. Reed, *Sunday Times*, 7 February 1971.

CHAPTER 10
1 *Scientific American*, July 1899, quoted by R. Dubos in *Reason Awake! Science for Man*.
2 F. M. Fisher et al. *Journal of Political Economy*, Vol. 70, 1962, p. 433.
3 *American Engineer*, April 1963, quoted in article by John Macinko in *Environment*, Vol. 12, No. 5, 1970, p. 8.
4 T. J. Chow, J. L. Earl, *Science*, Vol. 169, 1970, p. 577.
5 *Restoring the Quality of Our Environment*, Report of the Environmental Pollution Panel, President's Science Advisory Committee, The White House, Washington DC, 1965, p. 205.
6 *Chemical Engineering*, 4 May 1970.
7 *Chemical and Engineering News*, 10 February 1970.
8 *Chemical and Engineering News*, 16 March 1970.
9 *Chemical and Engineering News*, 16 February 1970.
10 *Chemical and Engineering News*, 22 July 1968, pp. 38-9.
11 Hohenemser & McCaull, *Environment*, Vol. 12, No. 5, 1970, p. 14.
12 *Chemical and Engineering News*, 22 July 1970.

CHAPTER 11
1 Historical material taken from J. Jewkes et al., *The Sources of Invention*, 1958, pp. 377-81.
2 J. I. Rodale et al., *Our Poisoned Earth and Sky*, 1969.
3 Mrs Lynn Miller, bacteriologist, Mary Manne College, Conference of American Water Works Association, 1961, quoted by Rodale.
4 *Chemical and Engineering News*, 2 September 1963.
5 *Chemical and Engineering News*, 18 March 1963.

6 Ibid. p. 110.
7 K. Mellanby, *Pesticides and Pollution*, 1967, p. 54.
8 A. F. Bartsch, Director, Pacific North West Water Laboratory and Chief, National Eutrophication Research Programme, FWPCA, for meeting at Lansing, Michigan, 10 February 1969.
9 *Chemical Engineering*, 1 June 1970, p. 71.
10 The contents of a Typical Heavy-Duty Granular-Detergent Formulation are:

Material	Per cent. wt.	Purpose
Sodium dodecylbenzene sulfonate	18	Surfactant
Sodium xylene sulfonate	3	Dedusting agent
Diethanolamide of coconut fatty acids	3	Foam booster
Sodium tripolyphosphate	50	Improves cleaning power
Sodium silicate	6	Anticorrosion agent
Carboxymethylcellulose	0.5	Soil redeposition preventative
Optical brightener	0.3	Fluorescent whitener
Benzotriazole	0.1	Antitarnishing agent
Other inorganic salts and water	19.1	Fillers
	100.0	

(Hearings, Conservation and Natural Resources Sub-committee, Committee on Government Operations, US House of Representatives, *Phosphates in Detergents and the Eutrophication of America's Waters*, 15 and 16 December 1969.)
11 *Chemical and Engineering News*, 27 April 1970, p. 14.
12 *Chemical and Engineering News*, 19 February 1968.
13 *Chemical and Engineering News*, 15 June 1970, p. 11.
14 A. L. Hammond, *Science*, Vol. 172, 1971, pp. 361-3.
15 S. E. Epstein, *Environment*, Vol. 12, No. 7, 1970, pp. 2-11.
16 L. M. Lichtenstein et al., *Journal of Allergy*, Vol. 47, No. 1, 1971, pp. 53-5.
17 *Chemical and Engineering News*, 22 June 1970, pp. 14-15.
18 *Chemical and Engineering News*, 3 February 1969, p. 17.
19 E. E. Angino et al., *Science*, Vol. 168, 1970, p. 389.
20 Quoted in 'The Integrity of Science', reprinted in Houlton, *Science and Culture*, 1965.

CHAPTER 12

1 R. J. S. Hookway, *The Management of the Environment in Tomorrow's Europe*, Report on Theme IV, Leisure, Council of Europe, 1970.
2 *Financial Times*, 22 August 1970.
3 Road Research Laboratory Reports, L.R. 288, L.R. 354.
4 Hookway, op. cit.
5 Roskill Commission Papers and Proceedings, Vol. 7, part 3.

CHAPTER 13

1 Quoted by P. A. Baran and P. M. Sweezy, *Monopoly Capital*, 1966, p. 210.
2 P. M. Boffey, *Science*, Vol. 168, 1970, p. 679.
3 The Half life is the time it takes for half the radioactive nuclei in a sample to emit their radiation. The Half lives of some of the principal isotopes produced by the fission of Uranium-235 and Plutonium-239 are:

Isotope	Half-life
Krypton-85	10 years
Rubidium-87	6.1×10^{10} years
Strontium-90	28 years
Ruthenium-106	about 10 months
Iodine-131	8.1 days
Cesium-137	33 years
Cerium-144	9 months
Europium-155	1.7 years

4 E. J. Kormondy, *Ecology and Man*, 1969.
5 M. McClintock, *Environmental Effects of Weapon Technology*, SIPI Workbook 1970.
6 M. Leitenberg, *Environment*, Vol. 12, No. 6, 1970, p. 26.
7 *Nature*, Vol. 226, 1970, p. 890.
8 G. R. N. Jones, *New Scientist*, 1971, pp. 690-91.
9 *The Times*, 14 November 1968.
10 *Chemical and Engineering News*, 26 May 1969.
11 W. Proxmire, *Report from Wasteland, America's Military Industrial Complex*, 1970.
12 *New York Post*, 22 September 1969.
13 V. Brodine, Workbook on *Environmental Effects of Weapons Technology*, SIPI 1970.
14 Teitelbaum, Workbook on *Environmental effects of Weapons Technology*, SIPI 1970.
15 Ibid.
16 *Science*, 25 September 1970, p. 1296.

17 *New York Times*, 23 August 1970.
18 BSSRS Newsheet No. 3, 1969.
19 *The Sunday Times*, 10 August 1969.
20 BSSRS Newsheet No. 3, 1969.
21 *Science*, 2 October 1970, p. 42.
22 G. H. Orians and E. W. Pfeiffer, *Science*, Vol. 168, 1970, pp. 544-54.
23 *Nature*, Vol. 228, 1970, p. 109.
24 Table. *Estimated area* treated with herbicides in Vietnam.*

Year	Defoliation (Acres)	Crop destruction (Acres)	Total (Acres)
1962	17,119	717	17,836
1963	34,517	297	34,814
1964	53,873	10,136	64,009
1965	94,726	49,637	144,363
1966	777,894	112,678	888,572
1967	1,486,446	222,312	1,707,758
1968	1,297,244	87,064	1,384,308

* Actual area sprayed is not known accurately because some areas are resprayed. (G. H. Orians and E. W. Pfeiffer, *Science*, Vol. 168, 1970, p. 544.)

25 *Composition Defoliant sprays used in Vietnam*

Agent	Composition	Use	% of total Defoliant sprayed
Orange	50-50 mixture of n-butylesters of 2,4 dichlorophenoxyacetic acid (2,4-d) and 2,4,5-trichlorophenoxyacetic acid (2,4,5-T)	General defoliation of forest, brush and crops	50%
Purple	50% n-butylester 2,4-D 30% n-butylester 2,4,5-T 20% 150 butylester 2,4,5-T	General defoliation interchangeable with orange	
White	2,4-D in amine formulation and picloram	Long term forest defoliation	35%
Blue	27.7% Sodium cacodylate 4.8% cacodylic acid	Rapid short term defoliation grass and rice	15%

(Table from *Scientific Research*, 9 June 1969, p. 27.)

26 F. M. Tschirley, G. M. Orians, E. W. Pfeiffer and the AAAS team led by M. S. Meselson.
27 G. M. Orians and E. W. Pfeiffer, *Scientific Research*, 9 June 1969, pp. 22-30, 29 June, pp. 26-30.
28 *Science*, Vol. 161, 1968, p. 253.
29 *Chemical and Engineering News*, 23 September 1968, p. 28.
30 The three were: Barry Commoner, Gerald Holton and M. Burr Steinbach, *Science*, Vol. 161, 19 July 1968, pp. 254-5.
31 *Science*, 24 April 1970.
32 *Nature*, Vol. 228, 1970, p. 109.
33 *Nature*, Vol. 226, 1970, p. 309.
34 *The Sunday Times*, 19 April 1970.
35 B. Nelson, *Science*, Vol. 166, 1970, pp. 977-9.
36 P. M. Boffey, *Science*, Vol. 171, 1971, pp. 43-7.

CHAPTER 14
1 First Annual Report of the Council on Environmental Quality, August 1970.
2 E. Aynsley, *New Scientist*, 29 October 1969.
3 V. J. Schaefer, AAAS Conference, Dallas, Texas, December 1968.
4 Dr Water Orr Roberts, AAAS Conference, Dallas, Texas, 1968.
5 C. F. Wurster, *Biological Conservation*, Vol. I, 1969, pp. 123-9.
6 R. L. Rudd, *Pesticides and the Living Landscape*, 1964.
7 C. F. Wurster, *Science*, Vol. 159, 1968, pp. 1474-5.
8 C. Manwell and C. M. A. Baker, *Molecular Biology and the Origin of Species*, 1970.
9 N. I. Sax, *Dangerous Properties of Industrial Chemicals*, 1968.
10 Definition of rem: the rad is a measure of radiation corresponding to the absorption of 100 ergs per gramme of tissue. The rem is a measure that includes an estimate of the biological effectiveness of different types of radiation. For the types of radiation considered here the two terms are roughly equivalent.
11 R. W. Halcomb, *Science*, Vol. 167, 1970, pp. 853-5.
12 A. Tamplin and J. W. Gofman, *Population Control through Nuclear Pollution*, 1970.

CHAPTER 15
1 'Chemistry and the Ocean', *Chemical and Engineering News*, Vol. 42, No. 22, 1964, pp. 3A-48A.
2 Joint IMCO/FAO/UNESCO/WMO Group of Experts on the Scientific Aspects of Marine Pollution, March 1969.
3 A. M. Rawn, 1st International Conference on Waste Disposal in the Sea, Pergamon Press, New York 1966, quoted by Hedgpeth.

4 A. Kaiser, Engineers Report on Waste Disposal in the San Francisco Area, quoted by Hedgpeth.
5 J. W. Hedgpeth, *Environment*, Vol. 12, No. 3, 1970.
6 *Disposal of Solid Toxic Wastes*, HMSO 1970, p. 68.
7 *Science*, Vol. 167, 1970, p. 363.
8 S. H. Fonselius, *Environment*, Vol. 12, No. 6, 1970.
9 *Taken For Granted*, Ministry of Housing and Local Government, London 1970.
10 *New Scientist*, Vol. 32, 1966, p. 612.
11 D. C. Holmes et al., *Nature*, Vol. 216, 1967, p. 227. A. V. Holden et al., *Nature*, Vol. 216, 1967, p. 1274.
12 R. W. Riseborough et al., *Nature*, Vol. 220, 1968, p. 1098.
13 C. F. Wurster and D. B. Wingate, *Science*, Vol. 159, 1968, pp. 979-81.
14 C. F. Wurster, *Biological Conservation*, Vol. 1, 1969, pp. 123-9.
15 G. G. Polikarpov, *Radioecology of Aquatic Organisms*, New York, 1966.
16 C. E. Nash, *Marine Pollution Bulletin*, Vol. 1, No. 1 and 2, 1970.
17 *Marine Pollution Bulletin*, Vol. 1, No. 6, 1970.
18 *Global Ocean Research*, Report of a Joint Working Party of the Advisory Committee on Marine Resources Research, the Scientific Committee on Oceanic Research, and the World Meteorological Organisation, Ponza and Rome, 1969.
19 The best hopes for new fishing grounds are: The Arabian Sea, the Caribbean and Gulf of Mexico, the Gulf of Alaska, the Indonesian and New Zealand Shelves, the Northwest African Shelf and the Gulf of Guinea, the Argentine and Chilean Shelves, the Antarctic Sea, as well as continental slopes down to 2,000 metres.
20 W. F. McIlhenny, *Chemical Engineering*, 7 November 1966, p. 253.

CHAPTER 16
1 *Man's Impact on the Global Environment: Assessment and Recommendations for Action*, 1970.
2 M. H. Horn et al., *Science*, Vol. 168, 1970, pp. 245-6.
3 J. D. George, *Marine Pollution Bulletin*, Vol. 1, No. 7, 1970.
4 R. B. Clark, *Marine Pollution Bulletin*, Vol. 1, No. 2, 1970.
5 *Chemical and Engineering News*, 17 February 1969.
6 Fu-Shiang Chia, *Marine Pollution Bulletin*, Vol. 1, No. 5, 1970.
7 W. F. McIlhenny, *Chemical Engineering*, 7 November 1966, pp. 249-50.

8 Commission on Marine Science, Engineering and Resources, 1969.

CHAPTER 17
1 V. Shkatov, *Voprosy ekonomiki*, No. 9, 1969.
2 *Izvestia*, 25 September 1970.
3 T. Khachaturov, *Voprosy ekonomiki*, No. 1, 1969.
4 M. I. Goldman, *Science*, Vol. 170, 1970, pp. 37-42.
5 A. G. Kasymov, *Marine Pollution Bulletin*, Vol. 1, No. 7, 1970, pp. 100-103.
6 V. Shkatov, op. cit.
7 Ibid.
8 *Komsomolskaya Pravda*, 16 March 1967.
9 *Komsomolskaya Pravda*, 3 November 1966.
10 V. Shkatov, op. cit.
11 T. Khachaturov, op. cit.
12 V. Shkatov, op. cit.
13 *Chemical Engineering*, 4 May 1970.

CHAPTER 18
1 'Environmental Engineering. A Guide to Industrial Pollution Control', *Chemical Engineering*, 27 April 1970.
2 *Chemical and Engineering News*, 29 June 1970, p. 43.
3 R. R. Grinstead, *Environment*, Vol. 12, No. 10, 1970.
4 R. C. Bostron and M. A. Sherif, *Nature*, Vol. 228, 1970, pp. 154-6.
5 *Chemical and Engineering News*, 6 April 1970, p. 38.
6 H. Mayhew, *London Labour and the London Poor*, 2 vols, 1851.
7 *Chemical Engineering*, 6 April 1970, p. 38.
8 J. H. Prescott and J. E. Browning, *Chemical Engineering*, 18 May 1970.
9 'Continental Water Pollution', *Science and Technology*, June 1969, p. 30.
10 J. F. Ferraro and R. A. Lerman, *Science and Technology*, June 1969.
11 *Chemical and Engineering News*, 23 February 1970, p. 23.
12 Royal Commission on Environmental Pollution, First Report, February 1971, Cmnd. 4585.
13 *The Sunday Times Business News*, 28 February 1971.
14 J. F. Ferraro and R. A. Lerman, op. cit.

CHAPTER 19
1 The section on English legal controls owes much to unpub-

lished reports by J. McLoughlin of the Pollution Research Unit, Manchester University.
2 Rivers (Prevention of Pollution) Act, S. 2 (1) (a).
3 L. J. Carter, *Science*, Vol. 166, 1969, pp. 1487-91.
4 L. J. Carter, *Science*, Vol. 166, 1969, pp. 1601-6.
5 *Science*, Vol. 166, 1969, pp. 200-201.
6 Report on a Draft European Convention on the Protection of Fresh Water against Pollution, Council of Europe, Doc. 2561, 12 May 1969, pp. 16-32.
7 *American Journal of International Law*, 182 (1941).
8 *American Journal of International Law*, 160 (1959).

CHAPTER 20
1 A. Gramsci, *The Modern Prince and Other Writings*, 1967.
2 K. Marx, *Capital*, Vol. 1, p. 541.
3 *Environmental Quality*, The First Annual Report of the Council on Environmental Quality, US Government Printing Office, August 1970.
4 R. G. Slater et al., *Ecology and Power*, 1970.
5 *Time* magazine, 2 February 1970.
6 S. Hays, *Conservation and the Gospel of Efficiency*, 1959, p. 270.
7 *Chemical Engineering*, 23 May 1966, p. 99.
8 *The Economics of Clean Water*, Federal Pollution Control Administration, March 1970.
9 *The Times*, 6 July 1970.
10 *Fortune*, February 1970, p. 118.
11 *Chemical Engineering*, 23 May 1966.
12 *The Times*, 6 July 1970.
13 *Chemical and Engineering News*, 9 February 1970, p. 78.
14 *Fortune*, February 1970, p. 118.
15 *Industry Week*, 5 December 1969, p. 27.
16 *Chemical and Engineering News*, 9 February 1970, p. 78.
17 *Financial Times*, 13 October 1970.
18 *Sunday Times*, 28 February 1971.
19 *International Management*, September 1970.
20 R. G. Slater et al., op. cit.
21 *Nature*, Vol. 226, 1970, p. 300.
22 *Sunday Times*, 7 February 1971.
23 *The Times*, 26 August 1970.
24 *Royal Commission on Environmental Pollution*, HMSO 1971.
25 Quoted by J. W. Batey, *Combustion Engineering*, Doc. 8458, 18 April 1968.
26 *Sunday Times*, 26 April 1970.

CHAPTER 21

1 T. Kotabinski, *Praxiology, An introduction to the science of efficient action*, 1965.

2 R. N. Grose, 'Some Problems in Economic Analysis of Environmental Policy Choices' in *Proceedings of Symposium on Human Ecology*, US Department of Health, Education and Welfare, 1968.

3 L. B. Lave and E. P. Seskin, *Science*, Vol. 169, 1970, pp. 723-32.

4 A. H. Strickland, *Journal of Applied Ecology*, Vol. 3, 1966, (supplement), pp. 3-13.

5 N. W. Moore, 'Experience with Pesticides and the Theory of Conservation', *Biological Conservation*, Vol. 1, 1969, pp. 201-207.

6 J. Adams, *Area*, Institute of British Geographers, No. 2, 1970, pp. 1-9.

7 *Chemical and Engineering News*, 23 February 1970.

8 *Chemical and Engineering News*, 10 January 1969.

9 *Chemical and Engineering News*, 9 February 1970, p. 81.

10 'Secrecy and Dissemination in Science and Technology', *Science*, Vol. 163, 1969, p. 787.

CHAPTER 22

1 Canon Montefiore, Edinburgh University Teach-in on Pollution, 6 March 1970.

2 T. Malthus, *An Essay on the Principle of Population* (1798), 1970, p. 75.

3 H. Myint, *The Economics of the Developing Countries*, 1964.

4 P. R. Ehrlich, *The Population Bomb*, 1968.

5 Mandel, *Marxist Economic Theory*, Vol. 1.

Selected Bibliography

The most widely useful books and articles are marked with an asterisk.

ABEY-WICKRAMA, I. et al. 1969 Mental Hospital Admissions and Aircraft Noise. *The Lancet*, 13th December, 1969, pp. 1275-1277.

*ALLEE, W. C., EMERSON, A. E., PARK, O., PARK, T. and SCHMIDT, K. P. 1949 *Principles of Animal Ecology*. Philadelphia and London. W. B. Saunders.

ARCHER, J. 1970 Effects of Population Density on Behaviour in Rodents. In: *Social Behaviour in Birds and Mammals*, ed. J. H. Crook. London and New York Academic Press.

ARGINO, E. E. et al. 1970 Arsenic in Detergents: possible danger and pollution hazard. *Science, 168*, pp. 389-390.

ARVILL, R. 1967 *Man and Environment: crisis and the strategy of choice*. London, Penguin Books.

AYNSLEY, E. 1969 How Air Pollution Alters Weather, *New Scientist*, 9th October, pp. 66-67.

*BARAN, P. A. and SWEEZY, P. M.   1966 *Monopoly Capital*. London, Penguin Books.
BELL, A.   1966 *Noise*. World Health Organisation.
BENTHEM, R. J.   1970 The Management of the Environment in Tomorrow's Europe. Report on Urban Conglomerations. *European Conservation Conference*, Strasbourg.
*BERNAL, J. D.   1969 *Science in History*. 4 Vols. London. Pelican.
BOFFEY, P. M.   1970 Hiroshima/Nagasaki. Atom Bomb Casualty Commission perseveres in sensitive studies. *Science, 168*, pp. 679-683.
BOFFEY, P. M.   1971 Herbicides in Vietnam: AAAS Study finds widespread devastation. *Science, 171*, pp. 43-47.
BOSTROM, R. C. and SHERIF, M. A.   1970 Disposal of Waste Material in Tectonic Sinks. *Nature, 228*, pp. 154-156.
BRYCE-SMITH, D.   1971 Lead Pollution—a growing hazard to public health. *Chemistry in Britain, 7(2)*, pp. 54-56.
CARTER, L. J.   1969 D.D.T.: the critics attempt to ban its use in Wisconsin. *Science, 163*, pp. 548-51.
CARTER, L. J.   1970 Nerve Gas Disposal: How the AEC refused to take the army off the hook. *Science, 169*, pp. 1296-1298.
*CHILDE, V. G.   1941 *Man Makes Himself*. London. Watts and Co.
CHILDE, V. G.   1950 The Urban Revolution. *The Town Planning Review, xxi*, pp. 3-17.
CHOW, T. J. and EARL, J. L.   1970 Lead Aerosols in The Atmosphere: Increasing Concentrations. *Science, 169*, pp. 577-580.
CHRISTENDON, L. D.   1963 New Approaches to Pest Control and Eradication. The Male Annihilation Technique in the Control of Fruit Flies. *Advances in Chemistry Series, 41*, pp. 31-35.
CLARK, R. B.   1970 Oil at Sea. *Marine Pollution Bulletin, 1(2)*, p. 1.
CLAUSEWITZ, K. von.   1968 *Clausewitz on War*. Ed: Rapoport, A. London. Pelican.
COLLEY, J. R. T. and REID, D. D.   1970 Urban and Social Origins of Childhood Bronchitis in England and Wales. *The British Medical Journal 2*, pp. 213-217.
CONNEY, A. H.   1967 Pharmacological Implications of Microsomal Enzyme Induction. *Pharmacology Review, 19*, pp. 317-366.
COURTNEY, K. D., et al.   1970 Teratogenic Evaluation of 2,4,5-T. *Science, 168*, pp. 864-866.
COX, J. L.   1970 DDT Residues in Marine Phytoplankton: increase from 1955-1969. *Science, 170*, pp. 71-73.
DAVIES, J. M.   1966 Epidemiology of Occupational Tumours of the Bladder. *Proceedings of the Royal Society of Medicine, 59*, pp. 1247-1248.

*DeBach, P. (ed) 1964 *Biological Control of Insect Pests and Weeds*. London. Chapman and Hall.
Dillard, D. 1950 *The Economics of John Maynard Keynes*. London. Crosby Lockwood.
Dinunno, J. J. and Levine, S. 1970 Environmental Matters Concerning Nuclear Electrical Power Production. *Nuclear Engineering International*, July/August, pp. 607-612.
Doll, R. et al. 1965 Mortality of Gasworkers with Special Reference to Cancers of the Lung and Bladder, Chronic Bronchitis, and Pneumoconiosis. *British Journal of Industrial Medicine*, 22, pp. 1-12.
Doxiadis, C. A. 1969 The City (ii). Ecumenopolis, World-wide City of Tomorrow. *Impact of Science on Society*. Vol. xix, (2), pp. 179-193.
Dubos, Rene 1970 *Reason Awake*: Science for Man. New York and London. Columbia University Press.
Egler, F. E. 1966 Pesticides in our Ecosystem. Communication II. *Bioscience*, *14*(ii), pp. 29-36.
Ehrlich, P. R. and Birch, L. C. 1967 The Balance of Nature and Population Control. *The American Naturalist*, *101*, pp. 97-107.
Ellis, H. M. 1967 Solid Waste Disposal. *European Committee for the Conservation of Natural Resources*. Strasbourg. Council of Europe.
Elton, C. S. 1942 *Voles, Mice and Lemmings, problems in population dynamics*. Oxford.
*Engels, F. 1954 *The Dialectics of Nature*. Moscow, Foreign Languages Publishing House.
Epstein, S. E. 1970 NTA. *Environment*, *12*(7), pp. 2-11.
Evans, F. C. 1956 Ecosystem as the Basic Unit in Ecology. *Science*, 123, pp. 1127-1128.
Ferraro, J. F. and Lerman, R. A. 1969 Pollution is a Dirty Word. *Science and Technology*, June, pp. 43-45.
Fisher, F. M., Griliches, Z., and Kaysen, C. 1962 The Costs of Automobile Changes Since 1949. *The Journal of Political Economy*, 70, pp. 433-451.
Flindt, M. L. H. 1969 Pulmonary Disease Due to Inhalation of Derivatives of *Bacillus subtilis* Containing Proteolytic Enzyme. *The Lancet*, 14th June, pp. 1177-1181.
Fonselius, S. H. 1970 Stagnant Sea. *Environment*, *12*(6), pp. 2-13.
Fu-Shaing Chia 1970 Reproduction of Arctic Marine Invertebrates. *Marine Pollution Bulletin*, *1*(5), pp. 78-79.
Galley, R. A. E. and Stevens, T. G. R. 1966 Pesticide Research Today. *Span*, *9*(2), pp. 107-109.

GEORGE, J. D. 1970 Sub-lethal Effects on Living Organisms. *Marine Pollution Bulletin*, *1*(7), pp. 107-109.
GLEN, R. 1954 Factors That Affect Insect Abundance. *Journal of Economic Entomology*, *47*, pp. 398-405.
GLORIG, A. 1961 The Problem of Noise in Industry. *American Journal of Public Health*, *51*, pp. 1338-1346.
*GOLDMAN, M. I. 1967 *Controlling Pollution. The Economics of a Cleaner America.* Englewood Cliffs, N.J. Prentice-Hall.
GOLDMAN, M. I. 1970 The Convergence of Environmental Disruption. *Science*, *170*, pp. 37-42.
GOLDSMITH, J. R. 1968 Effects of Air Pollution on Human Health. In: A. C. Stern, *Air Pollution*. Vol. 1.
GRAMSCI, A. 1967 *The Modern Prince and Other Writings*. London. Lawrence and Wishart.
GREENBERG, D. S. 1969 Pollution Control: Sweden sets up an ambitious new program. *Science*, *166*, pp. 200-201.
*GREGORY, P. 1965 *Polluted Homes*. London. Bell and Sons.
GRINSTEAD, R. R. 1970 The New Resource. *Environment*, *12*(10), pp. 2-17.
HAMMOND, A. L. 1971 Mercury in the environment: natural and human factors. *Science*, *171*, pp. 788-789.
HAYS, S. 1959 *Conservation and the Gospel of Efficiency*. Harvard University Press.
HEADLEY, J. C. and LEWIS, J. N. 1967 *The Pesticide Problem. An Economic Approach to Public Policy*. Resources for the Future Inc., Washington, D.C.
HEDGPETH, J. 1970 The Oceans: World Sump. *Environment*, *12*(3), p. 40.
HOHENEMSER, K. and MCCAULL, J. 1970 The Windup Car. *Environment*, *12*(5), pp. 14-21.
HOLCOMB, R. W. 1969 Oil in the Ecosystem. *Science*, *166*, pp. 204-206.
HOLCOMB, R. W. 1970 Radiation Risk: A scientific problem? *Science*, *167*, pp. 853-855.
HOLMES, D. C. et al. 1967 Chlorinated Hydrocarbons in British Wildlife. *Nature*, *216*, p. 227.
HOOKWAY, R. J. S. 1970 *European Conservation Conference.* The management of the environment in tomorrow's Europe. Report on Theme iv, Leisure. Council of Europe. Strasbourg.
HORN, M. H., TEAL, J. M., and BACKUS, R. H. 1970 Petroleum Lumps on the Surface of the Sea. *Science*, *168*, pp. 245-246.
HOURIHANE, D. O. B. 1964 The Pathology of Mesotheliomata and an Analysis of Their Association with Asbestos Exposure. *Thorax*, *19*, pp. 286-287.

*HEUPER, W. C. 1969 *Occupational and Environmental Cancers of the Urinary System*. New Haven and London. Yale University Press.
*HUTCHINSON, G. E. 1970 The Biosphere. *Scientific American*. 223, September.
JEWKES, J., et al. 1958 *The Sources of Invention*. Macmillan, London.
JONES, G. R. N. 1971 CS in the balance. *New Scientist*, 17 June, pp. 690-691.
JONES, M. G. and DAVIES, J. T. 1958 External Dust Deposition and Sulphur Emission. In: *Air and Water Pollution in the Iron and Steel Industry*. Special Report No. 61. Iron and Steel Institute, London. pp. 16-23.
*KAPP, K. W. 1963 *Social Casts of Business Enterprise*. Asia Publishing House.
KASYMOV, A. G. 1970 Industry and Productivity of the Caspian Sea. *Marine Pollution Bulletin*, *1*(7), pp. 100-103.
KNEESE, A. V., et al. 1970 *Economics and the Environment*. Resources for the Future, Inc., Washington, D.C.
KNIGHT, C. B. 1965 *Basic Concepts of Ecology*. London, Macmillan.
KNIPLING, E. F. 1962 Potentialities and Progress in the Development of Chemosterilants for Insect Control. *Journal of Economic Entomology*, 55, pp. 782-786.
*KORMONDY, E. J. 1969 *Concepts of Ecology*, Englewood Cliffs, N.J., Prentice-Hall Inc.
KOTABINSKI, T. 1965 *Praxiology*: An introduction to the science of efficient action. London. Pergamon Press.
KOEMAN, J. H. 1969 Chlorinated Biphenyls in Fish, Mussels and Birds from the River Rhine and Netherlands Coastal Area. *Nature*, *221*, pp. 1126-1128.
KOVDA, V. 1970 Contemporary Scientific Concepts Relating to the Biosphere. In: *Use and Conservation of the Biosphere*. Paris, UNESCO. pp. 13-29.
KUHN, T. S. 1962 *The Structure of Scientific Revolutions*. Chicago. University of Chicago Press.
LANDES, D. 1969 *The Unbound Prometheus: technological change and industrial development in Western Europe from 1750 to the present*. Cambridge University Press.
*LANGE, O. 1963 *Political Economy*. Vol. 1. London. Pergamon Press.
LAVE, L. B. and SESKIN, E. P. 1970 Air Pollution and Human Health. *Science*, *169*, pp. 723-732.

LEE, R. B. 1969 !Kung Bushman Subsistence: an input-output analysis. In: A. P. Vayda, *Environment and Cultural Behavior*.
LEFEBVRE, H. 1969 *The Explosion. Marxism and the French Upheaval*. New York, Monthly Review Press.
LEITENBERG, M. 1970 So Far, So Good. *Environment*, *12*(6), pp. 26-35.
LICHTENSTEIN, L. M. et al. 1971 Sensitization to Enzymes in Detergents. *Journal of Allergy*, *47*(1), pp. 53-55.
LINDEMAN, R. L. 1942 The Trophic-dynamic Aspect of Ecology. *Ecology*, *23* (4), pp. 399-417.
LÖFROTH, G. 1969 Man and DDT Compounds. *Deutsch-Schwedisches Symposium auf Gebilt der Umwesthygiene*, Baden-Baden.
LÖFROTH, G. 1970 Alkylation of DNA by Dichlorvos. *Die Naturwissenschaften*, *57*(8), pp. 393-394.
*LÖFROTH, G. 1970 *Methylmercury*. Ecological Research Committee. Bulletin No. 4. Swedish Natural Science Research Council. Stockholm.
LUNN, J. E., KNOWELDEN, J., and HANDYSIDE, A. J. 1967 Patterns of Respiratory Illness in Sheffield Infant Schoolchildren. *British Journal of Preventive and Social Medicine*, *21*, pp. 7-16.
MACINKO, J. 1970 The Tailpipe Problem. *Environment*, *12*(5), pp. 6-13.
MACLEOD, R. M. 1965-66 The Alkali Acts Administration, 1863-84. The Emergence of the Civil Scientist. *Victorian Studies*, *ix*, pp. 81-112.
MALDAGUE, M. E. 1970 Agriculture and Forests. *European Conservation Conference*. The management of the environment in tomorrow's Europe. Strasbourg. Council of Europe.
MALINOWSKI, B. 1944 *A Scientific Theory of Culture*. Chapel Hill. University of North Carolina Press.
MALTHUS, T. 1970 *An Essay on the Principle of Population* (1798). London. Penguin.
*MANDEL, E. 1968 *Marxist Economic Theory*. 2 Vols. London. Merlin Press.
MANHEIM, F. T., MEADE, R. H. and BOND, G. C. 1970 Suspended Matter in Surface Waters of the Atlantic Continental Margin from Cape Cod to the Florida Keys, *Science*, *167*, pp. 371-376.
MANWELL, C. and BAKER, C. M. A. 1970 *Molecular Biology and the Origin of Species*. London, Sidgwick and Jackson.
*MARX, K. 1924 *Capital*. Vol. 1. Chicago. Charles H. Kerr and Co.
MAYHEW, H. 1851 *London Labour and London Poor*. 2 Vols. London.
MCILHENNY, W. F. 1966 The Oceans: technology's new challenge. *Chemical Engineering*, 7th November, pp. 191-254.

McClintock, M. 1970 Environmental Effects of Weapons Technology. In: *Scientists Institute for Public Information Workbook.* New York.
*Mellanby, K. 1967 *Pesticides and pollution.* London, Collins.
*Miller, M. W. and Berg, G. G. 1969 *Chemical Fallout: Current research on persistent pesticides.* Springfield, Illinois. Charles C. Thomas.
Milne, A. 1962 On a Theory of Natural Control of Insect Population. *Journal of Theoretical Biology 3,* pp. 19-50.
Mishan, E. J. 1971 *Cost-benefit Analysis.* London. Unwin University Books.
Mitchell, H. W. 1965 Pesticides and other Agricultural Chemicals as a Public Health Problem. *American Journal of Public Health,* 55(7), pp. 10-15.
Moore, N. W. 1969 Experience with Pesticides and the Theory of Conservation. *Biological Conservation,* 1, pp. 201-207.
Morgan, K. Z. 1969 Tainted Radiation. *Science and Technology,* June, pp. 46-50.
Mumford, L. 1968 City: forms and functions. In the *International Encyclopedia of the Social Sciences.* New York. Macmillan. pp. 447-455.
Myrdal, G. 1968 *Asian Drama: An Inquiry into the Poverty of Nations.* London. Penguin. 3 Vols.
Nash, C. E. 1970 Marine Fish Farming. *Marine Pollution Bulletin,* *1*(1), pp. 5-6 and *1*(2), pp. 28-30.
Needham, J. 1964 Science and Society in East and West. In *The Science of Science.* Ed. Goldsmith M. and Mackay A. London. Souvenir Press. pp. 159-188.
Novick, S. 1970 The Burden of Proof. *Environment, 12*(8), pp. 16-29.
*Odum, E. P. 1959 *Fundamentals of Ecology.* Philadelphia and London, Saunders.
*Orians, G. H. and Pfeiffer, E. W. 1969 Mission to Vietnam. In 2 parts. *Scientific Research,* 9th June, pp. 22-30, 29th June, pp. 26-30.
Passino, I. R. 1970 The Management of the Environment in Tomorrow's Europe. Council of Europe report on industry. *European Conservation Conference.* Strasbourg.
Prescott, J. H. and Browning, J. E. 1970 Solid Waste Schemes Sifted. *Chemical Engineering,* 16th May.
Price, D. K. et al. 1968 On the Use of Herbicides in Vietnam—A Statement by the Board of Directors of the American Association for the Advancement of Science. *Science, 161,* pp. 253-254.

PROXMIRE, W. 1970 *Report from Wasteland. America's Military Industrial Complex.* Praeger.
RISEBOROUGH, R. W. et al. 1968 Polychlorinated Biphenyls in the Global Ecosystem. *Nature,* 220, pp. 1098-1102.
RODALE, J. I. 1969 *Our Poisoned Earth and Sky.* Rodale Books Inc. Pennsylvania.
*RUDD, R. L. 1964 *Pesticides and the Living Landscape.* Madison. University of Wisconsin Press.
SAX, N. I. 1968 *Dangerous Properties of Industrial Chemicals* (3rd ed.). New York, Reinhold.
SCOTT, T. S. 1963 Some Occupational Disease Hazards—10. Occupational Cancer of the Bladder. *Occupational Health,* xv(6), pp. 313-323.
SCOTT, T. S. 1965 Advances in Industrial Medicine. *The Practitioner, 195,* pp. 535-543.
SELIKOFF, I. J. 1967 Community Effects of Pneumoconiotic Dusts. In *Pneumoconiosis, a symposium.* World Health Organisation.
SIMMONDS, F. J. 1967 The Economics of Biological Control. *Journal of the Royal Society of the Arts, 115,* pp. 880-898.
SJOBERG, G. 1955 The Pre-Industrial City. *The American Journal of Sociology,* LX, pp. 438-445.
SJÖRS, HUGO 1955 Remarks on Ecosystems. *Svensk Botanisk Tidskift.* Bd *49* (H. 1-2), pp. 153-169.
SLATER, R. G. et al. 1970 *Ecology and Power.* San Francisco. People's Press.
SLONIM, N. and ESTRIDGE, N. K. 1969 Ozone, an Underestimated Environmental Hazard. *The Journal of Environmental Health, 31* (6).
*SMITH, R. F. and VAN DEN BOSCH, R. 1967 In: Kilgore, W. W. and Doutt, R. L. *Pest Control: biological, physical and selected chemical methods.* New York. Academic Press.
*STERN, A. C. (ed) 1968 *Air Pollution.* 3 Vols. New York, London. Academic Press. 2nd ed.
STOKINGER, H. E. and COFFIN, D. L. 1968 Biological Effects of Air Pollutants. In: A. C. Stern, *Air Pollution.* Vol. 1.
STRICKLAND, A. H. 1966 Some Estimates of Insecticide and Fungicide Usage in Agriculture and Horticulture in England and Wales, 1960-64. *Journal of Applied Ecology, 3* (supplement), pp. 3-13.
*TAMPLIN, A. and GOFMAN, J. W. 1970 *Population Control Through Nuclear Pollution.* Nelson Hall.
TANSLEY, A. G. 1935 The Use and Abuse of Vegetational Concepts and Terms. *Ecology 16* (3) pp. 248-307.
VAN DEN BOSCH, R. 1970 Pesticides: prescribing for the ecosystem. *Environment, 12*(3), pp. 20-25.

VAYDA, A. P. (ed)   1969 *Environment and Cultural Behaviour*. New York. The Natural History Press.
VEYS, C. A.   1969 Two Epidemiological Inquiries into the Incidence of Bladder Tumors in Industrial Workers. *Journal of the National Cancer Institute, 43*, pp. 219-226.
WALLER, R. E.   1952 The Benzpyrene Content of Town Air. *British Journal of Cancer, 6*, pp. 8-21.
WALTERS, H.   1970 Aspects of the use of Organic and Non-Organic Fertilisers. *Journal of the Soil Association, 16*(2), pp. 113-115.
WALTERS, H.   1970 Nitrate in Soil, Plants and Animals. *Journal of the Soil Association, 16*(3), pp. 149-170.
WOLMAN, A.   1967 The Metabolism of Cities. In: *Cities*, a Scientific American Book.
WRIGHT, G. W.   1969 Asbestos and Health in 1969. *American Review of Respiratory Disease, 100*(4).
WURSTER, C. F. and WINGATE, D. B.   1968 DDT Residues and Declining Reproduction in the Bermuda Petrel. *Science, 159*, pp. 979-981.
WURSTER, C. F.   1968 DDT Reduces Photosynthesis by Marine Plankton. *Science, 159*, pp. 1474-1475.
WURSTER, C. F.   1969 Chlorinated Hydrocarbon Insecticides and the World Ecosystem. *Biological Conservation, 1*, pp. 123-129.
WURSTER, C. F.   1969 Chlorinated Hydrocarbon Insecticides and Avian Reproduction: How are they related? In: Miller, M. W. and Berg, G. G. *Chemical Fallout; current research on persistent pesticides*. Springfield, Illinois. Thomas. pp. 368-389.
YARWOOD, C. E.   1970 Man-made Plant Diseases. *Science, 168*, pp. 218-220.
—— *The Prevention and Control of Dust Diseases in Industry*. British Chemical Industry Safety Council. London.
—— 1954 *Proceedings of the 4th European Seminar for Sanitary Engineers*. Opatija, Yugoslavia.
—— 1963 Committee on the Problem of Noise. *Noise: Final Report*. London. Her Majesty's Stationery Office.
—— 1963 FAO Symposium on the nature and extent of water pollution. Problems affecting inland fisheries in Europe. Jablonna, Poland.
—— 1964 US Congress: Committee on Government Operations: Sub-committee on Reorganisation and International Organisations: Inter-agency Co-ordination in Environmental Hazards, Pesticides, Washington, D.C. 1963-4 (The 'Ribicoff' Committee Hearings).
—— 1964 European Conference on Air Pollution. Council of Europe. Strasbourg.

―― 1964 *Occupational Diseases in California Attributed to Pesticides and other Agricultural Chemicals.* Californian State Department of Health.

―― 1964 *Review of Persistent Organochlorine Insecticides.* London. Her Majesty's Stationery Office.

*―― *The Integrity of Science.* A report by the American Association for the Advancement of Science Committee on Science in the Promotion of Human Welfare. In: Holton, G. (1965). Science and Culture. Boston. Houghton Mifflin Co.

*―― 1965 *Restoring the Quality of our Environment.* Report of the Environmental Pollution Panel. President's Science Advisory Committee. The White House, Washington, D.C.

―― 1966 *Fresh Water Pollution Control in Europe.* Council of Europe. Strasbourg.

―― 1966 WHO Technical Report, No. 318. WHO Expert Committee on Water Pollution Control.

―― 1967 *Air Pollution Engineering Manual.* US Dept. of Health Education and Welfare. National Centre for Air Pollution Control. Washington, D.C.

*―― 1967 *Cities.* A Scientific American Book. London. Penguin Books.

―― 1967 *Power Systems for Electric Vehicles.* A Symposium. US Dept. Health, Education and Welfare. No. 999-AP-37. Washington, D.C.

―― 1967 Report on an Inter-Regional Seminar on Water Pollution Control. New Delhi, India. World Health Organisation.

―― 1968 *Air Pollution Control.* Report on an inter-regional seminar convened by the World Health Organisation in collaboration with the Government of the USSR.

―― 1968 *Noise-Sound Without Value.* Federal Council for Science and Technology. Washington, D.C.

―― 1968 Proceedings of Symposium on Human Ecology. US Dept. of Health, Education and Welfare. Washington, D.C.

*―― 1969 *Cleaning Our Environment. The Chemical Basis for Action.* Washington, D.C. American Chemical Society.

―― 1969 *Global Ocean Research,* Report of a Joint Working Party of the Advisory Committee on Marine Resources Research, the Scientific Committee on Oceanic Research, and the World Meteorological Organisation. Ponza and Rome.

―― 1969 Joint IMCO/FAO/UNESCO/WHO Group of Experts on the Scientific Aspects of Marine Pollution. London.

―― 1969 *Poison in our Air.* United Steelworkers of America National Conference on Air Pollution. In co-operation with National Air Pollution Control Administration. Washington, D.C.

―― 1969 Report on a draft European convention on the protection of fresh water against pollution. Strasbourg. Council of Europe. Doc. 2561.

*―― 1969 *Resources and Man*. Committee on Resources and Man. National Academy of Sciences―National Research Council. San Francisco. W. H. Freeman and Co.

―― 1969 Secrecy and Dissemination in Science and Technology. A Report of the Committee on Science in the Promotion of Human Welfare. *Science, 163*, pp. 787-790.

―― 1969 *Solid Waste Processing: A state of the art report*. US Department of Health, Education and Welfare. Washington, D.C.

―― 1970 *Air Quality Criteria for Carbon Monoxide*. Summary and Conclusions. US Dept. of Health, Education and Welfare. Washington, D.C.

―― 1970 *Air Quality Criteria for Hydrocarbons*. Summary and Conclusions. US Dept. of Health, Education and Welfare. Washington, D.C.

―― 1970 *Air Quality Criteria for Photochemical Oxidants*. Summary and Conclusions. US Dept. of Health, Education and Welfare. Washington, D.C.

―― 1970 *Air Quality Criteria for Particulate Matter*. Summary and Conclusions. US Dept. of Health, Education and Welfare. Washington, D.C.

―― 1970 *Air Quality Criteria for Sulfur Oxides*. Summary and Conclusions. US Dept. of Health, Education and Welfare. Washington, D.C.

―― 1970 *Clean Water for the 1970s*. US Department of the Interior. Federal Water Quality Administration. Washington, D.C.

―― 1970 Commission ('Roskill') on the Third London Airport. Papers and proceedings. *vii*, part 3. London, Her Majesty's Stationery Office.

―― 1970 *Disposal of Solid Toxic Wastes*. London. Her Majesty's Stationery Office.

―― 1970 *The Economics of Clean Water*. Federal Pollution Control Administration. Washington.

*―― 1970 Environmental Engineering: A guide to industrial pollution control. *Chemical Engineering* (Deskbook Issue), 27th April.

*―― 1970 *Environmental Quality*. The first annual report of the Council on Environmental Quality, Washington, D.C.

*―― 1970 *Man's Impact on the Global Environment: Assessment and Recommendations for Action*. Report of the Study of Critical Environmental Problems (SCEP). Cambridge, Mass., MIT.

—— 1970 *The Mercury Pollution Problem in Michigan and the Lower Great Lakes Area.* Michigan Water Resources Commission.

—— 1970 *Modern Farming and the Soil.* Report of the Agricultural Advisory Council on Soil Structure and Soil Fertility. London. Her Majesty's Stationery Office.

—— 1970 *The Price of Convenience.* The Committee for Environmental Information. *Environment,* 12(8), pp. 2-15.

—— 1970 *The Protection of the Environment.* London. Her Majesty's Stationery Office. Cmnd. 4373.

—— 1970 Summary of report on pollution of Lakes Erie and Ontario and international section of St Lawrence River. NATO letter, xviii/2.

*—— 1970 *Taken for Granted.* Ministry of Housing and Local Government. London. Her Majesty's Stationery Office.

*—— 1970 *Use and Conservation of the Biosphere.* Paris, UNESCO.

—— 1970 *Vibration Syndrome.* Interim report by the Industrial Injuries Advisory Council. London. Her Majesty's Stationery Office. Cmnd., 4430, June, 1970.

—— 1970 Water pollution problems and control programs in Michigan of the Great Lakes. Michigan Water Resources Commission.

*—— 1971 Royal Commission on Environmental Pollution. First Report. London. Her Majesty's Stationery Office. Cmnd. 4585.

Index

Abel, I. W., 71
Advertising, 303-4, 321
Agriculture, 18, 34-5; Health Hazards to Workers, 82-4; Economic Trends in, 91-2; Declining Labour Force, 91-92; Intensified Exploitation of Land and Man, 92; and Food Processing Industry, 92-3
Air: Composition of, 48, 341; Biological Function of, 49
Air Pollution, 48-60, 251; Primary Pollutants of, 48; and Disease, 50; and Lead, 61 151; and Diet 75; and Dust 75; and Asbestos 76; Effects on Global Climate 205-9; The Greenhouse Effect, 206-7; by Particles, 206-208; Control Techniques, 262; Legislation Relating to, 280-1, 284-5; Bronchitis, 52-3; and Social Class, 72-5; Lung Cancer, 53, 54, 341, 342
Air Travel, Scale of, 179
Airports, 69; Cost of Benefit Analysis of Sites, 314
Aldrin: Banning of, 101, 102
Alkali Acts (UK), 281, 294
American Association for the Advancement of Science, on Use of Defoliants in Vietnam, 198, 202; on Secrecy, 317-18
Angino, E. E., 172
Arctic, Sensitivity to Pollution, 243-5
Armaments Industry, 182-183
Arsenic, 53; in Detergents, 172-3; in Baltic Sea, 224
Atomic Bomb Casualty Commission (ABCC), 185-6
Atomic Energy Commission (US), 217
Austria, Water Pollution, 120, 121
Authoritarianism, 326, 333
Automobiles, Scrap, 64, 65; Industry, 148, 149; as A Pollution Source, 150, 151; Unconventional Power Sources, 156-7; Pollution Control Measures, 155, 156; Effects on Countryside, 178
Automobile Emissions, Carbon Monoxide, 56; Nitrogen Dioxide, 58, 59; and Photochemical Smog, 59; Lead, 61

Balance of Nature, 8, 16-17, 18, 340
Baltic Sea, Pollution of, 223-5
Bartsch, A. F., 164
Benzpyrene, 53-5, 64; In Sea, 240
Beryllium, 53
Beta-Napthylamine, 79-81
Biological Warfare, 194-195
Biosphere, 5-6, 19
Birds, and Organo-Chlorine Insecticides, 100-101; Mercury in, 137-141, And DDD, 210, 231; and PCB, 227, 228; and Oil Pollution, 240-241, 243; and Insecticides, 295, 314
Birth Control, 329-30
Bourgeoisie, The, 39
British Society for Social Responsibility in Science, 194
Brothel, 36
Bruin, P., 96
Bryce-Smith, D., 61-2
Bulgaria, Water Pollution in, 130-1
Business Cycle, Effect on Pollution, 321
Baker, C. M. A., 210

Cadmium, Pollution, 146-147; Disease (ITAI-ITAI), 146
Capitalistic Society, Development of, 38-42; and the Rise of Rationality, 41-2
Carbon, 14, 187
Carbon Dioxide, and Climatic Change, 206-7
Carbon Monoxide, 46, 55-57, 150, 342
Carson, R., 101
Case, R., 79-80
Caspian Sea, Pollution of, 250-1
Cesium, 137, 187, 214
City, and Pollution, 36-7, 45-7
Cities, Migration to, 47; Disease in, 47; Air Pollution 48, 54; Water Pollution in, 62-4; Public Health Problems in, 63
Cities, Pre-Industrial, 35-38
Childe, G., 27

Clark, R. B., 240
Clean Air Act (UK), 52, 283
Climate, 204-8
Climax Community, 17-18
Coleridge, S. T., 120
Colley, J. R. T., 74
Commodity Production, 38
Commoner, B., 123, 317
Commonwealth Institute of Biological Control, 111
Communal Succession, 17-18
Communism, Primitive, 30-31
Composting, 266-7
Consciousness, 20-1, 23-5
Cost Benefit Analysis, 311-316; of Pesticides, 109-112, 313-14; of Air Pollution Control, 313; of Airport Sites, 314
Cousteau, J., 223, 246
Cronin, E. L., 235
C. S. Gas., 189, 190

Darwin, C., 334
Davies, J. M., 80
DDT, 320, 328; Discovery of, 98; Persistence of, 99-100; in Mothers Milk, 100; Enzyme Induction by, 100; and Birds, 100-101, 210, 231; Banning of, 102; Global Distribution, 209; in Baltic Sea, 224; and PCB, 227; in Phytoplankton, 230-1; in Bermuda Petrel, 231
Defence Budgets, 182
De-Growth School, 325-6
Detergents, Enzyme, 170-2; Effect on Workers, 88; and Eutrophication, 125; Industry, 159-60; Synthetic, Development of, 160; as Water Pollutants, 161-6, Biodegradable, 162; and Phosphate Pollution, 164-8, Alternatives to Phosphates, 168-170; Nitrilotriacetate (NTA), 168-9; Alkylbenzene Sulphonates (ABC), 160-2; Soap, as an alternative to, 170; Arsenic in, 172-3; Composition of, 347
Developing Countries, 320; Water Pollution, 128-32;
Stages of Development, 128-9
Dichlorvos, 104-6
Dieldrin, Banning of, 101, 102
Dillard, D., 25
Doan, H. P., 300, 302
Dole, H. M., 269
Dubridge, L. F., 200

Earthday, 297
Ecology, 6-8, 11, 102; Social Significance of, 333, 335
Economics, 28, 29; and Pest Control, 109-12, of Lead Free Petrol, 152-155; Speculation in Pollution Control Stocks, 270-3; Cost Benefit Analysis, 311, 316; No-growth School, 325
Ecosystem, 8-13, 17-18, 335; Agricultural, 18; and the City, 46
Ecumenopolis, 47
Education, 287, 336, 338
Egypt, Pollution of Nile, 130
Ehrlich, R., 138
Elton, C., 8
Energy Flows, 10-13
Enforcement of Controls, 306-8
Engels, F., 21, 22, 23, 24
Enzyme Induction, 100
Environmental Defence Fund, 103, 286
Epstein, S. S., 169
Eutrophication, 125; in Lake Erie, 125, 127; and Detergents, 164-8
Evolution, 4-5, 20

Faith, N., 308
Federal Radiation Council, 216
Fertilisers, 94-7; Production of, 95; as Cause of Methemoglobinaemia, 95-96; and Pests, 96-7
Fimreite, N., 141, 142, 144
Fire, 27
Firket, J., 51
Fish, Effects of Pollution on, 117, 120, 126, 251; Effects of Water Pollution on, 132; Effects of Mercury Pollution on, 135-7; Mercury in, 140, 142-5; Farming of, 233-

Fish—cont.
235; Potential New Fishing Grounds, 351
Flindt, M. L. H., 88
Fluorine, 307
Fluoroacetamide, 94
Food Chains, 22, 229; Mercury in, 144; Organochlorine Insecticides, Biological Concentration in, 209-10; Radioactive Materials in, 186, 187
Fog Disasters, 50-2, Meuse Valley, 51; Donora, 51; Poza Rica, 51; London, 51-2
Food, Social Surplus of, 29, 30, 33-6
Foster, J. S., 199
France, 47; Water Pollution, 120, 121
Franklin, B., 27
Frazer, J., 27
Freedom, 338
Fungicides, Methyl Mercury a Source of Pollution, 137

Galston, A. W., 202
Germany, Water Pollution, 119, 120; Detergents, 160, 162
Gladkov, N., 252
Global Problems, 204; Climatic, 204-8; Organochlorine Insecticides, 209-211; Radiation, 211-17
Gofman, J. W., 216
Great Barrier Reef, 243
Greece, Pollution of Beaches, 176
Greenhouse Effect, 206-7
Green Revolution, 22

Haeckel, E., 7
Hayes, S., 297
Herbicides, Use in Vietnam, 196-203, 349; Composition of Vietnam Defoliants, 197, 349; 2, 4-D, Toxicity of, 198-9; 2, 4, 5-T, Toxicity of, 199, 200, 201, 202; Cacodylic Acid, Toxicity of, 200; Teratogenic Effects of, 2, 4, 5-T and 2, 4-D, 200-202; Contaminated by Dioxin, 200-201
Hiroshima, Casualties, 185, 186
Holdo, J., 299, 302

Hourihane, D. O. B., 78
Hueper, W. C., 81, 295
Hungary, Water Pollution, 121; Bans Organo-Chlorine Insecticides, 102
Hunting and Gathering Economy, 27-31
Hydrocarbons, 46, 150
Hyerdahl, T., 223

Ideology, 332-3
Incinerators, 265
India, Water Pollution in, 129-30; Pollution and Disease in, 66
Industrialists, Attitudes to Pollution, 298-306
Industry, and the Military, 191-2
International Controls on Pollution, 288-90, 315, 337
Italy, Pollution of Beaches, 177
Industrial Accidents, 72
Insecticides, Revolution in, 97-8; Controversy over use of, 98-103; Alternatives to, 107-12; Costs of Research and Development, 109-10
Iodine-131, 186-7, 214
Ireland, F. E., 306
Irrigation, 94

James, R., 190
Japan, Water Pollution in, 131-2; Mercury Pollution in, 135-7; Cadmium Pollution, 146-7; Radiation Pollution, 185-186
Jenson, S., 227, 228
Jet Aircraft, and Climatic Change, 206, 208
Johnson, J., 201

Kato, S., 132
Khachaturov, T., 249
Knipling, E. F., 108
Knowledge, Flow of, 211, 310, 316-19
Kormondy, E. J., 8
Kovda, V., 5
Kraus, E. J., 196
Krypton-85, 214, 215
Kuhn, T., 7
Kung Bushmen, 28-30, 32
Labour, Division of, 28, 29, 33, 34, 36, 40; in

Pre-Industrial Cities, 37, 38
Labour Process, 25-7
Labour Theory of Value, 25
Lake Baikal, Pollution of, 249-50
Lake Erie, Pollution of, 124-8
Lapland, Radio Active Fallout, 187
Lave, L. B., 313
Lee, R., 28-30
Lead, 60-2; in Petrol, 61, 151-5; Poisoning in Children, 61-2; in Housepaint, 62
Legislation, 274-90; English, 277-83; on Oil Pollution, 238, 281-3; United States, 284-7; Swedish, 287-8; International, 288-290; Enforcement of Controls, 306-8; Factory, 72, 81-2, 84, 86-7, 88; Pesticides, 103; Lead Levels in Petrol, 153; Detergents and, 162, 163, 165, 166; Title to Marine Resources, 236; Pressure Groups, 286; Political Forces for Change in, 294-6
Leitenberg, M., 189
Lindeman, R. N., 10
Löfroth, G., 139

Malinowski, B., 27
Mallet, L., 240
Malthus, T., 326, 327, 329, 331
Markert, C. L., 202
Matter, Cycling of, 11-13, 340
Maximum Permitted Levels of Contaminants in Food, 140-1
Manwell, C., 210
Marx, K., 21, 25, 26, 27, 29, 296; On Industrial Health Legislation 72
Mayhew, H., 268
McDuffie, B., 144
Megalopolis, 47
Metha, R. S., 129
Mellanby, K., 164
Mercury, Toxic Action of, 136, 140-1, 143-4; Conversion Inorganic Mercury to Organic Mercury, 138-9; Maximum Permit-

Mercury—*cont.*
  ted Level in Fish, 140-141; Production of, 142
Mercury Pollution, 135-7; Minimata Disaster, 135-137; Sources of, 139-40; In sea, 144-5
Metals, Discovery of, 35
Miettinen, J. K., 187
Milne, A. J., 16
  Bed, 245-6
Minamata Disaster, 135-7
Mineral Resources, on Sea
Mitchell, H., 84
Molybdenum, 53
Moore, N., 314
Morality and Pollution, 324
Morgan, L. H., 32
Multi-National Corporations, 184
Mumford, L., 36

National Union of Agricultural and Allied Workers (UK), 83
Nature, Control of, 20-4, 25-6
Nature, Dialectical View of, 20-22
Needs, 27
Nelson, G., 297
Neolithic Revolution, 30-1, 34
Netherlands, 47; Water Pollution, 119, 120
Nitrilotriacetic Acid (NTA), 168-9
Nitrogen Dioxide, 58, 59
Nitrogen Oxides, 46, 150
Nixon, R., 296, 297
Noise, 66-70; Levels of, 67, 68; and Health, 68, 69; and Airports, 69; Costs of Soundproofing, 70; at work, 85-7; Control Technique, 262; Legislation, 283
Nuclear Explosions, Civil Use of, 214-15
Nuclear Power Stations, Windscale Incident, 210; Reactor Failures, 212; Siting of, 212; Disposal of Radio Active Waste, 213-14; Increasing Numbers of, 217; Legislation Relating to, 280
Nuclear Weapons, Pollution by, 184-9; Underground Tests of, 187-8, Partial Test Ban Treaty,

187; Accidents to, 188-9
Nutrient Cycles, 10-13, 340

Occupational Diseases, 76-89; and Dust, 76; Byssinosis, 76, 77; Asbestosis, 77, 78; Cancer of the Bladder, 78-82; in Agriculture, 82-4; Tunnel Workers Disease, 84-5;
Occupational Deafness, 85-7; and Technological Innovation, 87-9; Enzyme Detergents, 88; Vibration Syndrome, 88-89, 344
Occupational Health and Safety Act, (US), 72
Oceanography, 219
Oil, Pollution by, 237-45; Composition of, 238; Toxicity of Components, 239-40; Carcinogens in, 240; Pollution and Birds, 240-1, 243; Tankers and Pollution, 238, 242-3; Alaska, 243-5; Pollution of Caspian Sea, 250-1, Legislative Controls on Pollution, 281-2
Oil Dispersants, Toxicity of, 239, 241, 242
Organisation, Levels of, 6, 7
Organochlorine Insecticides, *See* Insecticides and DDT; Global Contamination by, 209-11
Orians, G. H., 197, 198, 200
Ozone, 59, 60

Particles, 206-8
Passino, I. R., 47
Pavlov, I., 338
'People' Pollution, 16
Permament Arms Economy, 184
Peroxyacetyl Nitrates (PAN), 59
Pest Control, 18; Integrated, 99, 107; Alternatives to Insecticides, 107-12; Costs and Benefits of Research and Development, 109-12; Sterile Male Technique, 108; Biological Control, 111
Pesticides, Hazards to Users, 83, 84; Scale of Use, in California, 83;

as a cause of Pollution, 97-112; Production of, 97; Costs of Development, 109-10; in Seas, 230-2; in Soviet Union, 252
Petroleum Industry, 152, 154, 155
Pfieffer, E. W., 197, 198, 200
Phosphates, in Detergents, 164-8
Photochemical Smog, 59
Pickett, A. D., 107
Plankton, 220; DDT in, 230-1
Planning, 335-8
Plant Diseases, Increasing Number of, 93
Pleasure Boats, Pollution by, 178, 179
Plowshare Project, 213-15
Poison Gas, 191; Problems of Disposal, 192-4
Poland, Water Pollution, 121
Politicians and Pollution, 296-8
Polychlorinated Biphenyls (PCB), 227-9; in Baltic Sea, 224
Pollution Control Technology, For Water Pollution, 260-1; Reverse Osmosis, 261; For Air Pollution, 262; for Noise, 262; for Smells, 263; for Solid Waste, 263; Incinerators, 265; Pyrolysis, 265-6; Composting, 266-267; Business Speculation in, 270-3; Expenditure on, 270, 296, 298-9
Population, Dynamics of, 14-18; and Behaviour, 15-16; Population Growth, 326-31, Causes of, 328, Rates of, 328, 329; Overpopulation Concept, 330
Potts, I. E., 266
Pre-Industrial Cities, Production in, 37, 38; Pollution of, 37
Production, Forces of, 26-27, 33, 34
Production, Primitive, 27-31
Production, Process of, 26
Progress, 23, 24
Profit, Concept of, 42; and

Profit—*cont.*
 Workers Health, 87, 89; and Pesticides. 103, 104; and Nuclear Power Industry, 217; Priority of, 299, 300, 302, 305; and Economic Rationality, 310
Proxmire, W., 191
Pyrolysis, 265-6

Radiation Damage, 185-7, 215-16, 217; from X-Rays, 216; Maximum Permissible Exposure Level, 216
Radioactive Materials, Half Life, 348
Ramazzini, B., 82, 85
Rationality, Economic, x, 41, 42, 309-10, 333; Social xi, 335, 337, 338; Impeded by Secrecy, 316-19
Reagan, R., 243
Recycling of Waste, 264-70
Refuse, 46
Reid, D. D., 74
Respiratory System, 49, 50
Reuse, H. S., 161, 191
Revolutionary Movement, 333
Ribant, J. P., 47
Ribicoff, A., 102
Riseborough, R. W., 228
Rodale, J. I., 171
Rohde, G., 96 97
Roosevelt T. 297
Rose S. 195 201
Roueche, B., 51
Royal Commission on Pollution (UK), 272, 306
Russell, W. T., 50

Santa Barbara, 243
Science, Unity and Fragmentation, 6, 7; and Ideology, 330-3; Paradigm, 7
Scientists, Social Responsibility of, 102; 'Silent', 102; Pressures on 102-3
Scott, T. S., 80
Seas, Pollution of, 218-19; Composition of 219-21; Major Sources of Pollution 221-3; as a Receptacle for Wastes 222-3; Pollution by Industrial Wastes 226; Pollution by PCB 227-9; Mine Waste Pollution by, 229, 230; Pollution by Pesticides, 230-2; Pollution by Radioactive Materials, 229; Thermal Pollution, 233; as a Source of Food, 233, 235-6; Pollution by Oil, 237-45; Undersea Mining, 245-6
Secrecy, 316-19
Second Homes, 179-80
Seskin, E. P., 313
Sewage, 46; and Marine Pollution, 225-6; Health Hazards of, 226
Sewage Systems, Ancient Rome, 37
Sevin, 110
Simmonds, F. J., 111
Social Classes, 34, 35
Social Crisis, 334
Socialist Society, 24, 338
Soil, 93-7; Fertility of, 94-95; and Chemical Fertilisers, 94-5; and Compost, 266
Solid Waste, 64-6, 263-70; and Air Pollution, 65; and Public Health, 66; Control Technique, 263; Salvage of, 267-70; Legislation, 283
Soviet Union, Benzpyrene Pollution in, 54, 55; Water Pollution in, 121, 248-51; Pollution of, 248-55; Russian Revolution, 248; Lake Baikal, Pollution of, 249-50; Caspian Sea, Pollution of, 250-1; Pollution, Effects on Fishing. 251; Air Pollution in, 251; Pesticides in, 252; Squandering of Natural Resources, 252-253; Bureaucracy and, 254-5
Sombart, W., 41, 42
Sternglass, E., 215, 216
Stewart, J., 25
Strontium-90, 186, 214, 317
Suess, E., 5
Sulphur Dioxide, 46, 57, 58; Control of, 262
Sun, 3-4
Sweden, Pesticide Legislation, 102; Mercury Pollution in, 137-41; Legislative Controls in, 287-8
Switzerland, Water Pollution, 119, 120

Tamplin, A., 216

Tansley, A. G., 9
Technology, x, 23, 24, 27, 30, 331, 332, 336; Unintended Effects of, 24
Technological Innovation, 87-9, 321, 322
Technological Optimism, 19
Teitelbaum, S. L., 193
Teller, E., 214
Thermal Pollution, 221, 233
Tools, 27-8
Torrey Canyon, 240, 241, 242
Tourism, Industry, 174-6; Mediterranean, 176-7
Toxicity, and Air Pollution, 50; and Behaviour, 58, 239
Traditionalism, 41
Tritium, 215
Trophic Levels, 12
Typhoid, 63

Unemployment, Pollution and, 307-8
United Kingdom, Water Pollution, 120, 121; Pesticides in, 101-2; Detergents, 162; Pollution of Beaches, 177; Car Ownership, 178; Legislative Controls in, 277-83
United Nations, 337
United States, Water Pollution, 121-8; Detergents, 162; Legislative Controls in, 284-7
U.S. Department of Defense, on Use of Defoliants in Vietnam, 199, 202
Urban Factor, in Disease, 54
Urban Revolution, 35, 36

Veblen, T., 320
Vernadsky, V., 5
Vietnam War, Ecological Damage from, 195-203; Birth Defects Resulting From, 202; Numbers of Bomb Craters, 198; Extent of Herbicide Spraying, 349
Voronova, L., 252

Waller, R. E., 54
Water, Biological function, 113, 114; Availability of, 114-16; Increasing Demands for, 115-16

Water Pollution, and Pathogenic Organisms, 63; Defined, 116-18; Principal Sources of, 117-18; Extent of Fresh Water Pollution, 118-32; in Europe, 119-21; in Britain, 121; in Soviet Union, 121; in United States, 121-8; in Great Lakes, 122-8; in Developing Countries, 128-32; Effects of, 132-4; Industrial Attitudes to, 132; by Detergents, 161-173; by Arsenic, 172-3; Legislation, 277-9, 284

Wilderness Areas, 180

Working Class, The, 39, 314, 315, 338; and Pollution, 70-89

Wurster, C., 209, 210, 231

X-Rays, 216

Yannacone, V., 103